Laboratory Experiments

for

World of Chemistry

Steven S. Zumdahl
Susan L. Zumdahl
Donald J. DeCoste

BROOKS/COLE

CENGAGE Learning™

Australia • Brazil • Japan • Korea • Mexico • Singapore • Spain • United Kingdom • United States

Laboratory Experiments for World of Chemistry
Steven S. Zumdahl, Susan L. Zumdahl, and
Donald J. DeCoste

For product information and technology assistance, contact us at
Cengage Learning Customer & Sales Support, 1-800-354-9706

For permission to use material from this text or product,
submit all requests online at **www.cengage.com/permissions**
Further permissions questions can be emailed to
permissionrequest@cengage.com

ISBN-13: 978-0-618-82967-5

ISBN-10: 0-618-82967-9

Brooks/Cole
20 Davis Drive
Belmont, CA 94002
USA

Cengage Learning is a leading provider of customized learning solutions with office locations around the globe, including Singapore, the United Kingdom, Australia, Mexico, Brazil, and Japan. Locate your local office at **international.cengage.com/region**

Cengage Learning products are represented in Canada by Nelson Education, Ltd.

To learn more about Brooks/Cole, visit **www.cengage.com/brookscole**

Purchase any of our products at your local college store or at our preferred online store **www.ichapters.com**

Printed in the United States of America
7 8 9 10 11   16 15 14 13

CONTENTS

Preface

This manual is intended for students who have never taken a chemistry course before. The experiments are designed to be challenging but understandable for the student. Experiments begin with basic laboratory techniques and observations, and progress to relatively complex procedures. We have included many more experiments than needed for a year course, in order to allow flexibility to meet student needs, and to provide experiments in a variety of formats. Our main objectives are to provide students with:

1. an introduction to laboratory experimentation
2. improved understanding of measurements and their limitations
3. familiarity with a variety of chemical reactions and the equations used to describe them
4. opportunities to collect and process data in traditional form, microscale format, and by computer-integrated systems
5. practice in analyzing and drawing conclusions from their own or class data

We have designed experiments in three different formats to illustrate the different ways that scientists collect and use information. **Traditional** experiments use beakers, flasks and traditional equipment to carry out experiments on a macroscale level. **Microscale** experiments provide opportunities for students to view chemical reactions on a small scale while cutting costs and reducing the effect on the waste stream. **CBL** experiments allow students to connect probes to calculators or computers and collect data electronically. These experiments illustrate the power of technology in the laboratory today and give students an opportunity to analyze their data and results electronically.

We have worked to provide a helpful and convenient format for both students and teachers. Each experiment includes:

1. a statement of the problem to be studied in the experiment
2. a discussion of the important principles in each experiment
3. detailed procedures to permit students to work relatively independently under supervision
4. names, and wherever practical, formulas of all reagents needed for the experiment in the **Materials** section
5. safety precautions in a special **Safety** section for each experiment, as well as safety notes in the **Procedure** and **Cleaning Up** sections
6. a section in each experiment guiding students through the cleaning up process and providing directions for waste management
7. an **Analysis and Conclusions** section which assists the students in thinking through the appropriate calculations and questions to understand the experiment

Each experiment has a **Report Sheet** in the Student Report Sheet book, which includes **Prelaboratory Questions**, a **Data/Observations** section, **Analysis and Conclusions** questions, **Summary Tables** and graph paper to assist students in writing their reports in an organized and logical manner. These report sheets are also designed to improve efficiency for teachers as they evaluate their students' work.

We are especially indebted to our colleagues John Little and Connie Grosse for their invaluable contributions to the writing of these experiments. Tim Loftus from University of Illinois spent many hours testing and evaluating experiments to improve them and to provide sample data for the Teacher's Guide. We appreciate the comments and suggestions of teachers and students from across the country that have provided us with many ideas. We would be happy to hear from you as you use the experiments. Please let us know your suggestions for improving the manual.

S. Zumdahl
S. Zumdahl
D. DeCoste

To The Student

The experiments in this laboratory manual are designed to help you learn chemistry by exploring the chemical nature of matter. Chemistry is an experimental science—we learn things by making careful observations and then formulating possible explanations for these observations. The primary purpose of the experiments in this manual is to help you to discover how the scientific method works. We also hope you will learn to appreciate the fascinating chemical world around us.

To provide you with a variety of laboratory experiences, we have included three different types of investigations, which you will recognize by the little icon at the top right of the first page of each experiment. The ones headed by a flask (⚗) are called **traditional** labs and use standard laboratory apparatus and glassware. Other experiments are done in **microscale**. As you can infer from the name, these experiments use very small, specialized equipment, including plastic test plates and pipettes. You can recognize them by the icon showing small test tubes and a magnifying glass: ⚗. The third category of experiments give you a chance to use digital electronic probes connected to a computer or to your graphing calculator. These technology-based activities are identified by a picture of a graphing calculator 🖩.

The experiments are organized to present the problem to be solved, to give you some background material necessary to understand the experiment, to help you analyze your results and to think beyond the present experiments to other related explorations. We will now consider how each of these goals is accomplished.

The **Problem** presents the question to be answered in the experiment. It is your job to be able to solve the problem through the experimental investigation. The **Introduction** gives you the essential background you need to understand the goals and methods of the experiment and often refers you to the text where more detailed information is available. The **Prelaboratory Assignment** contains questions that you should complete before you carry out the experiment. It is very important that you are well prepared before you begin the experiment so you can better appreciate what you are seeing and so you can carry out the experiment safely. The **Materials** list tells you what chemicals and other materials are needed for the experiment. These will be furnished by your teacher except in the cases in which you are requested to bring something from home. The **Safety** section calls your attention to potential hazards present in the experiment. Notice that you are always required to wear safety goggles and a lab apron (or lab coat) whenever you are working in the lab. By paying careful attention to the safety issues you can complete the experiment without injury to yourself or those around you. The **Procedure** gives an organized plan for carrying out the experiment. It is very important that you read and understand the procedure before you start working. In the course of the procedure you will make various observations for later use as you analyze your results. Sometimes the observations are qualitative—they require a description in words of what you see. Other observations are quantitative—they are measurements, and you will need to record numbers with the appropriate units. Be sure to record your observations as they occur. You will not remember them accurately later on. In the **Analysis and Conclusions** section you are asked questions that will help you understand how your observations are related to the problem you are trying to solve. You need to analyze your observations and then decide what they mean. The **Something Extra** gives you ideas of how your investigations can be extended in some way—some additional ideas you can explore.

Many of the experiments in this manual use traditional apparatus and techniques—beakers, flasks, and test tubes. Other experiments are done in microscale. These experiments use small amounts of chemicals and containers, such as well plates and plastic pipets, to illustrate chemical reactions and other phenomena. Finally, there are experiments that employ electronic devices hooked to calculators or computers to collect and analyze data. All of these experiments illustrate the ways that chemists explore the world around them.

Graphing Guidelines

A graph is a representation of the relationship between two physical quantities, such as temperature and time. Graphs summarize data. Because of this, graphs are used in sciences to show experimental results, physical relationships, trend, experimental conclusions and predictions.

Graphing involves several steps:
1. Collecting and organizing data
2. Plotting data
3. Drawing the graph
4. Interpreting the graph

A graph should be constructed so that it can stand by itself as a representation of the data. Therefore, each graph should include:

1. A **title** that identifies the data without reference to the lab or experiment number
2. Clearly **labeled axes** with units of calibration given. Each axis should be labeled in two ways, first with the measurement being graphed (e.g. time, temperature, volume, etc.) and second with units for the measurement (seconds, hours, °C, K, mL, L).
3. A reasonable **scale** should be chosen for the axes so that the graph is easy to read, or use for interpolation or extrapolation. The scale should be appropriate for the size of the paper. The graph should fill between half and a full page. The scale used on the horizontal axis does not need to be the same as the scale for the vertical axis.
4. Clear data points are necessary for producing a readable graph. Sometimes a small circle around the points can make them more visible.
5. When drawing your graph, be sure to draw the best **smooth line** that fits your data points. Do not "connect the dots" with straight segments. The "best fit line" may not actually pass through each data point, but it should come as close as possible to all of the points while remaining a smooth curve.
6. Use a sharp pencil or fine colored pen to create **sharp lines.** Use different colors to show contrast when more than one line appears on a particular graph. The overall appearance of your graph should be neat.

Working with Chemicals

Success in the chemistry laboratory depends on your confidence in handling materials and chemicals as you work. Knowing how to safely dispense and transfer chemicals, as well as how to use laboratory equipment, creates a safe work environment for everyone.

Measuring Liquids

How you measure the volume of liquid is determined by the precision needed in the measurement. Most beakers and flasks have graduations on the side similar to a measuring cup. These graduations are usually only accurate to about ±5%. If you need a more exact volume, use a graduated cylinder or a pipet. Choose the size for the measuring container that best matches the volume you need to measure. In graduated cylinders or burets, liquids often form a meniscus (a curved surface). For most liquids you should read the volume indicated at the bottom of the meniscus.

Transferring Liquids

The first problem encountered when preparing to measure the volume of a liquid is how to get it out of the reagent bottle without contaminating your sample or the liquid remaining in the bottle. If the bottle has a stopper a glass stopper, remove it in the following way: with your palm upward, grasp the top of the stopper between two fingers and remove it. Keep holding the stopper while you pick up the bottle and poor the liquid with the same hand. Replace the stopper. Your teacher will demonstrate the correct technique for you in class. This process prevents the stopper from ever touching another surface (besides the reagent bottle), preventing contamination of the stopper or your work area.

To prevent splashing when pouring a liquid into an open container (such as a beaker), place a stirring rod in the container and pour the liquid down the rod. When pouring a liquid into a small-mouthed container (such as a test tube), place the liquid in a beaker first. Then pour from the beaker into the test tube using the spout of the beaker. This will give you more control over the stream of liquid. Place the test tube in a test tube rack before you pour liquid into it to prevent spilling the liquid on your fingers.

Transferring Solids

When transferring a solid to wide-mouthed container, you can use a spoon, spatula or scoopula to collect and transfer the solid. When placing a solid into a small-mouthed container (such as a test tube) first crease a piece of paper (about 4 –6 cm square) in half. Then place the solid onto the paper in the crease. Pour the solid into the small container from the creased paper. Waxed paper works best for this.

Mixing Chemicals

Be sure that you use a container large enough to hold all the chemicals with extra room for mixing. Never use a thermometer to stir mixtures of chemicals. Use a stirring rod, magnetic stirrer, wooden split, or toothpick instead.

When mixing chemicals in a test tube, put a stopper on the test tube and place your thumb over the stopper. Shake the tube up and down, holding it so that the mouth of the test tube points away from you and others. Remove the stopper slowly in case mixing has resulted in the build up of pressure in the test tube.

Using a Laboratory Balance

Chemists frequently use a balance to measure mass. Although balances differ considerably in style, and you may have more than one type in your classroom, they all work on the same basic principle.

One of the most common balances consists of a suspended pan for holding the material to be weighed and a set of calibrated beams with riders. The object is placed on the pan, and the riders are moved until the pan is centered. Then the mass is read. You may have encountered this type of balance in your doctor's office.

The balance is a sensitive instrument and must be handled carefully. Do not carry balances from place to place in the classroom. You should go to the balance to measure mass. Check your balance to see that it is calibrated correctly. If the pointer swings evenly above and below the midpoint with no object on the pan, the calibration is correct. If you have a balance that is out of calibration, contact your teacher for assistance in recalibrating it, Be sure you read the steps for using a balance and practice several times with familiar objects until you are comfortable with the procedure.

1. Set all the riders to zero, and remove any material from the balance pan.
2. Single solid objects may be placed directly on the pan. Liquids or granular solids (such as salt or sugar) should be placed in a pre-weighed container or on weighing paper.
3. Move the largest rider along the beam one notch at a time. When the pointer drops, move the rider back one notch. Repeat this procedure with each smaller rise until you finally use the smallest rider.
4. The beam for the smallest rider has no notches. Slowly slide this rider along the beam until the pointer is swinging equally on either side of the midpoint. You may want to move the rider with a pencil point to keep the disturbance of the balance to a minimum.
5. The mass of the object on the pan is the sum of the masses of all the riders. Be sure to subtract the mass of the container or weighing paper if you used either.

The precision of a balance depends on the smallest division of the scale. Most quadruple beam balances have scales with divisions to 0.01 g. Since you can estimate between these marks, the precision of the balance is 0.001 g.

Electronic balances are easier to use and are becoming more common in high school classrooms. To use an electronic balance, place the container or weighing paper on the pan, press the zero (or tare) button, and wait until the reading is 0. Next add the material you wish to measure. The balance will give the measure mass a digital readout. Notice that in the case of an electronic balance the balance estimates the last digit in the readout.

Be sure to leave the balance in clean condition when you have finished your measurement. Take all used containers, weighing papers, and excess reagent back to your work area or dispose of them properly. Chemicals should never be placed directly on the balance pan for weighing.

Common Laboratory Equipment

Bunsen burner

Pipestem triangle

Evaporating dish

Test tubes

Beaker

Utility clamp

Ring stand

Iron ring

Mortar and pestle

Crucible and cover

Gas bottle

Safety goggles

Corks

Watch glass

Erlenmeyer flask

Wire gauze

Tongs

Assorted rubber stoppers

Pipet

Test tube holder

Lab burner

Thermometer

Dropper

Forceps

File

Wire brush

Buret

Graduated cylinder

Wash bottle

Micropipets (standard and narrow stem)

Spatula

Funnel

Test-tube rack

24-well plate

Scoopula

Experiment 1

Chemistry of Fire

Problem

What is the most efficient flame for a Bunsen burner? How is a Bunsen burner used to heat various substances?

Introduction

The Bunsen burner is the most common source of heat in the chemistry laboratory. Natural gas enters the burner near the base, rises through a barrel, mixes with air, and burns. By proper adjustments the most efficient flame can be obtained.

The typical Bunsen burner contains the parts shown in **Figure 1**. Variations such as a combination collar and barrel are often seen. Look at your burner and identify each of the parts shown in **Figure 1**. Gather the materials given in the list below and begin your investigation of the problem.

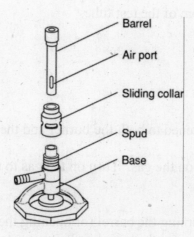

Figure 1
Parts of a Bunsen Burner

Prelaboratory Assignment

✓ Read the **Introduction** and **Procedure** before you begin.
✓ Answer the Prelaboratory Questions.
  1. Draw a Bunsen burner and label its parts.
  2. What is the proper color for a burner flame?

Materials

Safety goggles	Bunsen burner
Lab apron	Wire gauze
Matches	Test tube and holder
Tongs	Ring stand and iron ring
Copper wire	Metric ruler
Stirring rod	100-mL graduated cylinder
Thermometer	250-mL beaker
Hot pad	

Safety

1. Wear safety goggles and lab apron whenever working in the laboratory.
2. When using a Bunsen burner:
   - ✓ Confine long hair and loose clothing.
   - ✓ Never leave a burner unattended.
   - ✓ Do not reach over the burner.
   - ✓ Make sure no flammables are near the burner.
3. When heating test tubes:
   - ✓ Do not point the open end toward any person.
   - ✓ Do not heat the bottom of the test tube.

Procedure

Part 1   Use of the Burner

Setting the flame

1. Be sure the hose is securely fastened to both the burner and the gas outlet.

2. Strike the match *before* turning on the gas. Turn on the gas to the maximum open position and light the burner.

3. The flame can be adjusted by turning the barrel (or opening the air ports).
   a. If the flame is a yellow color, use the barrel and air ports to adjust it until it is completely blue and shows an inner cone, as illustrated in **Figure 2**.

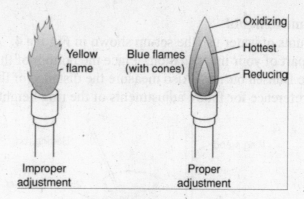

Figure 2
Adjusting a burner flame

b. If the flame is blue without the inner cone, adjust the barrel or air ports until an inner cone is visible.

The hottest part of the flame

4. Use tongs to hold a piece of copper wire and insert it into the flame just above the top of the barrel.

5. Slowly lift the wire up through the flame. Observe and record the color of the copper wire at various heights. The wire is hottest when it glows red. Record the position of the hottest part of the flame.

6. Remove the wire from the flame and allow it to cool on a heat resistant surface.

Heating a test tube

7. Half fill a test tube with tap water. Apply heat to the side of the test tube as shown in **Figure 3**. Notice that the top of the test tube is at an angle and pointed away from all people in the area.

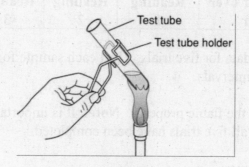

Figure 3
Heating a liquid in a test tube

8. If heat is applied directly to the test tube, the contents will shoot out of the tube or "bump". A slow smooth movement of the tube in and out of the flame should be used to heat the contents evenly. Once the water is boiling, remove the test tube from the flame, discard the water into the sink and set the tube and holder aside to cool.

Heating a larger volume of liquid

9. To heat larger amounts of water use the set-up shown in **Figure 4**. Assemble this apparatus now for use in the next part of your investigation. Place the bottom of the beaker exactly 15.0 cm above the top of the Bunsen burner. Also measure the distance of the ring from the surface of the lab bench as a reference for future adjustments of the ring height.

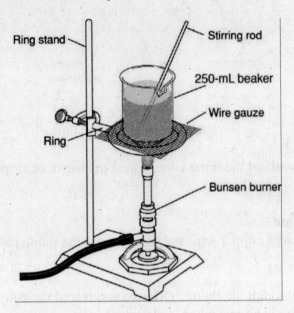

Ring stand

Stirring rod

250-mL beaker

Wire gauze

Ring

Bunsen burner

Figure 4
Heating a larger amount of liquid

Part 2   Efficiency of the Bunsen Burner

If you are not using a Report Sheet for this experiment make a Data Table for this section with the headings shown below:

Trial	Volume of water	Height of beaker over burner	Temp Reading 1	Temp Reading 2	Temp Reading 3	Temp Reading 4	Temp Reading 5

1. Use the table to record the data for five trials. Heat each sample for five minutes, taking the temperature at one-minute intervals.

2. Light the burner and adjust the flame properly. **Note:** It is important that the burner not be turned off or changed until all five trials have been completed.

3. For each trial do the following:
   a. Measure 80 mL of water using a graduated cylinder and pour it into the beaker.
   b. Place the burner under the beaker and heat the water for five minutes taking the temperature of the water every minute.

c. For each temperature reading:
   1. The water should be stirred with a glass stirring rod just before each reading.
   2. The thermometer should not be used to stir the water, nor should it be allowed to rest on the bottom of the beaker.
   3. The temperature will change very rapidly when the thermometer is first placed in the water. Wait to record the temperature until the rate of change is slow.
   4. Remove the thermometer and carefully place it on the table. Thermometers break very easily! Do not shake these thermometers or wave them in the air.
   d. After the last temperature reading, remove the burner from under the beaker. Remember not to turn it off!

4. After each trial is complete, carefully remove the beaker (use tongs or hot pad) from the stand and pour the water into the sink. Thoroughly rinse the beaker with cool water to bring it back to room temperature.

5. Lower the ring about 2.5 cm from the previous setting. **Be careful, the ring is hot**! Record the new height accurately on your data table.

6. Repeat steps 2 – 5 for each remaining trial. Be sure to record the data after each measurement.

Cleaning Up

1. Any broken glass should be carefully placed in a labeled container for broken glass.
2. Carefully clean your glassware and return all of your equipment to its proper location.
3. Wash your hands before leaving the laboratory.

Analysis and Conclusions

Complete the **Analysis and Conclusions** section for this experiment either on your Report Sheet or in your lab report as directed by your teacher.

1. Make a graph of temperature versus time. Plot time on the *x*-axis (horizontal) and temperature on the *y*-axis (vertical). Place the plots for all of the trials on the same graph. Label each line with the height of the beaker above the burner.

2. Which line shows the most efficient heating of the water? Explain why you chose this line.

3. What are the advantages of using a blue flame instead of a yellow one for heating objects in the laboratory?

4. Where is the hottest part of the blue flame?

5. When heating a substance over a Bunsen burner where should the object be placed for most efficient heating? Why?

6. How does graphing the data help to determine the most efficient height for heating a liquid in a beaker?

Experiment 2

Scientific Observations

Problem
What observations indicate that a chemical reaction has occurred? What is the difference between simple physical mixing of ingredients and an actual chemical reaction?

Introduction
You may already feel that you are a good observer, but there is more to good observation than you might think. Observing scientifically requires concentration, attention to detail, the effective use of measuring tools, ingenuity and patience. It also takes practice.

Before you come to lab, try making scientific observations at home. Find a candle and light it. See how many observations you can write down for the burning candle. Michael Faraday, a 19th century chemist, found many such things to observe in a burning candle. In fact, he wrote a book on the subject and gave talks about it.

Compare the observations you made about your candle to those your teacher gives you in class. Those observations that are descriptive are called *qualitative*.   Observations that involve numerical quantities are *quantitative* – these are called measurements. Chemists often measure temperature, time, mass, or volume.

Now you can practice your skills of scientific observation by watching what happens when several substances are mixed together.

Prelaboratory Assignment
- ✓ Read the **Introduction** and **Procedure** before you begin.
- ✓ Make observations of a burning candle. (Never leave a burning candle unattended.)
- ✓ Answer the Prelaboratory Questions.

    1. Which of the following statements is *not* a scientific observation? For those that are not observations, briefly explain why they are not.
        a. The wire is 18 cm long.
        b. The solution is blue-green in color.
        c. The reaction is producing a red-brown solid. It must be rust.
        d. The gas is being produced at a rate of 6.5 mL/second.

    2. What distinguishes an observation from a conclusion?

Materials

Safety goggles
Lab apron
Plastic teaspoon
Thermometer

Small beaker (100-mL)
Blue-green chemical
Aluminum foil (15 cm x 15 cm)
Glass stirring rod

Safety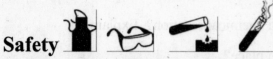

1. Wear safety goggles and a lab apron at all times in the laboratory.
2. No food or drink is allowed in the laboratory at any time.
3. Observe odors by gently waving the fumes toward your face. Do not sniff directly from a container.
4. Unknown chemicals can be caustic. Avoid contact with your skin.
5. If you are using a mercury thermometer, be very careful. Mercury is extremely toxic. If you break a mercury thermometer, report it to your teacher immediately.

Procedure

Note: In this experiment you may use the thermometer at any time. No specific directions for its use will be given.

1. Fill the small beaker half-full with tap water. Add one rounded spoonful of the blue-green chemical to the water **gently, without stirring**.

2. Write your observations of what is happening at each step of the procedure as a list.

3. Note the appearance of the solution in the beaker. Stir the chemical into the water gently. Record any observations.

4. Form the piece of aluminum foil into a loosely rolled tube about the size of a hot dog. Stand the aluminum tube in the solution in your beaker and wait several minutes. Record your observations of what happens.

5. When you are convinced that the reaction is finished, remove the remaining aluminum foil, if any, from the beaker and place it in the trash or solid waste disposal container provided by your teacher.

Cleaning Up

1. Carefully pour the solution that remains in the beaker into the proper disposal container. Do not pour any solid into this container.
2. Place any remaining solid from the bottom of the beaker in the solid waste container.
3. Thoroughly wash your hands before leaving the laboratory.

Analysis and Conclusions

Complete the **Analysis and Conclusions** section for this experiment either on your Report Sheet or in your lab report as directed by your teacher.

1. Do your observations from Procedure, step 1 indicate a chemical reaction? Explain.

2. Do your observations from Procedure, step 3 indicate a chemical reaction? Explain.

3. Do your observations from Procedure, step 4 indicate a chemical reaction? Explain.

4. What are some clues that a chemical reaction has taken place?

5. Identify the three qualitative observations you made that you think are most noticeable. Were any of your observations quantitative? If so, which?

6. Classify each of your observations as qualitative or quantitative.

7. Is dissolving a solid in water a chemical reaction? Why or why not?

8. Is a metal disappearing into a solution a chemical reaction? Why or why not?

Something Extra

1. Design an experiment that would separate the blue-green chemical from the water.

2. What name is given to chemical reactions that produce energy as heat? What are those reactions that consume energy as heat called?

Experiment 3

Observations and Explanations

Problem
Why does a flame burn? How does an observation differ from a theory?

Introduction
Observation is an important part of science, but it is often misunderstood by beginners. Many students believe that "observing" is the same as "seeing". However, observing is much more than seeing. For example, at the scene of an accident different witnesses will often give different testimony as to what they saw. The witnesses all saw the same thing, but some are more observant than others.

This lab consists of a simple system – a burning candle. You have undoubtedly *seen* a burning candle many times before. In this lab, *observe* the burning candle.

Prelaboratory Assignment
✓ Read the entire experiment before you begin.

Materials
Apparatus
Safety goggles
Lab apron
Candle
Drinking glass or beaker (taller than the candle)
Watch or clock (with a second hand)
Matches
3" x 5" index card

Safety
1. You are working with a flame in this lab. Tie back hair and loose clothing.
2. Do not drop matches into the sink. Dispose of burned matches in the trashcan after they are cooled.
3. Safety goggles and a lab apron must be worn at all times in the laboratory.

Procedure
Part 1
1. Light a candle.
2. Drip some wax onto a 3" x 5" card and stand the candle in the melted wax. Hold the candle upright until it can stand alone.
3. Observe the burning candle for a few minutes.

Part 2

1. Place a drinking glass upside down over the candle.

2. Observe the candle for a few minutes. Record the time it takes for the flame to *almost* go out.

3. Lift the glass from the candle and place it on the table (mouth down).

4. Replace the glass over the candle. Record the time it takes for the flame to go out completely.

Part 3

1. Remove the glass and place it mouth side down on the table.

2. Light the candle again.

3. Lift the glass from the table and lower the mouth of the glass over the burning candle.

4. Observe what happens.

Cleaning Up

1. Clean up all materials and wash your hands thoroughly.
2. Dispose of all chemicals as instructed by your teacher.

Data/Observations

Complete the **Data/Observations** section for this experiment either on your Report Sheet or in your lab report as directed by your teacher.

Part 1

1. Make a list of observations of the burning candle.

Part 2

2. What happens to the flame when the candle is covered?
3. How long does it take for the flame to almost go out after the candle is covered?
4. What happens to the flame when you lift the glass?
5. How long does it take for the flame to go out after the candle is covered again?

Part 3

6. Do you observe anything on the inside of the glass? Discuss.
7. What happens to the flame when you lower the glass over the candle?

Analysis and Conclusions

Complete the **Analysis and Conclusions** section for this experiment either on your Report Sheet or in your lab report as directed by your teacher.

1. Compare your list of observations from Part 1 with another group of students. Which observations did you make that they did not? Which observations did they make that you did not?

2. Compare the times from questions 3 and 5 in the Data/Observations section.

3. Develop a theory that explains your observations when the candle is covered with the glass. Make sure to address the following:
   a. Support your theory with specific observations.
   b. How does your theory explain the differences in times when covering the candle (in Part 2)?
   c. No theories answer all questions. What are two questions that your theory does not answer or address?

4. Why does a flame burn? Explain how your observations support your answer.

5. What is the difference between a theory and an observation? Give an example of each from this experiment.

Something Extra

A few hundred years ago (before oxygen was discovered) scientists proposed a theory of burning that relied on a substance called phlogiston. In this theory, substances that burned were said to contain phlogiston. Burning resulted in the release of phlogiston, a substance that could not burn. Thus, the flame of a burning candle placed under a glass will eventually go out because the glass will become filled with phlogiston. Design an experiment to disprove the theory of phlogiston.

Experiment 4

Physical and Chemical Changes

Problem
How do physical changes differ from chemical changes?

Introduction
Matter can undergo physical and chemical changes. Physical changes involve change of one or more physical properties. Chemical changes involve changes in the composition of the substance.

The following are clues that indicate that a chemical change may have occurred: a color change, the evolution of bubbles, the formation of a solid, and heat absorbed or produced.

In this lab you will carry out four procedures and make observations. Then you will decide how well the clues enable you to determine if a chemical reaction occurred. In addition, you will develop molecular definitions of physical and chemical changes.

Prelaboratory Assignment
✓ Read the entire experiment before you begin.

Materials

Apparatus	Reagents
Safety goggles	Barium nitrate
Lab apron	Potassium sulfate
250-mL beakers (2)	HCl (3.0 M)
100-mL beakers (3)	Magnesium ribbon
Graduated cylinder	Food coloring
Bunsen burner	Water
Matches	
Ring stand and ring	
Wire gauze	
Spatula	
Sponge	
Well plate (6 wells)	
Tongs	

Safety
1. The 3.0 M HCl is corrosive. Handle it with extreme care.
2. If you come in contact with any solution, wash the contacted area thoroughly.
3. You are working with a flame in this lab. Tie back hair and loose clothing.
4. Do not drop matches into the sink. Dispose of burned matches in the trashcan after they are cool.
5. Safety goggles and a lab apron must be worn at all times in the laboratory.

Procedure

Part 1

1. Add about 100 mL of water to a 250-mL beaker.

2. Add a few drops of food coloring to a beaker of water.

3. Make and record observations for several minutes.

Part 2

1. Dissolve a spatula tip amount of barium nitrate in a small amount of water (about 10 mL) in a 100-mL beaker.

2. Dissolve a spatula tip amount of potassium sulfate in a small amount of water (about 10 mL) in a 100-mL beaker.

3. Pour the solutions from Step 1 and 2 together into an empty 100-mL beaker. Make and record careful observations.

Part 3

1. Add about 100 mL of water to a 250-mL beaker.

2. Arrange the beaker and ring stand as shown in **Figure 1**.

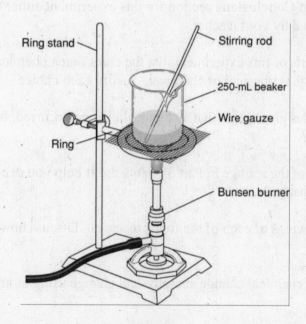

Ring stand
Stirring rod
250-mL beaker
Wire gauze
Ring
Bunsen burner

Figure 1
Apparatus to boil water

3. Light the Bunsen burner and place it under the beaker. Adjust the burner so the hottest part of the flame touches the bottom of the beaker.

4. Bring the water to a boil. Make and record careful observations.

5. Use the tongs to hold the sponge over the beaker for a couple of minutes as the water boils.

6. Place the sponge on the lab bench and let it cool.

7. Squeeze the sponge. Record your observations.

Part 4
1. Place a few drops of 3.0 *M* HCl in one of the wells of your well-plate.

2. Add a small piece of magnesium ribbon to the acid. Make and record careful observations.

Cleaning Up
1. Clean up all materials.
2. Wash your hands thoroughly.
3. Dispose of all chemicals as instructed by your teacher.

Analysis and Conclusions
Complete the **Analysis and Conclusions** section for this experiment either on your Report Sheet or in your lab report as directed by your teacher.

1. For each of the four parts of this experiment, list the clues that a chemical change occurred, and tell whether each change is physical or chemical. Justify each choice.

2. Do any of the procedures give a clue that a chemical change occurred, but are not chemical changes? Which ones?

3. What was the purpose of the sponge in Part 3? How did it help you decide if the process was a physical or chemical change?

4. Make microscopic drawings of each of the four processes. Discuss how these explain your observations.

5. Develop definitions of chemical change and physical change using atoms and molecules in your definitions.

Something Extra
Chemists have learned that a chemical change always includes a rearrangement of the ways in which atoms are grouped. Explain what this statement means and discuss whether your observations support this statement.

Experiment 5

Household Chemistry

Problem

How can you identify a common household chemical by comparing its properties with the properties of known materials?

Introduction

We use chemicals everyday to cook, bake, and sometimes to feel better. The chemical and physical properties of a substance make up an identification tag for each substance. A chemist can use these properties to identify an unknown substance or to determine the contents of a mixture of substances.

When substances are mixed together two things can happen:
- ✓ Nothing. The substances are mixed but no change occurs.
- ✓ Formation of new substances. The substances interact to produce one or more different substances. This process is called a chemical reaction.

How do you know if a chemical reaction has occurred? Important clues that a chemical change has happened are: a color change, the production of a gas, the formation of a solid, or the production of heat.

In this experiment you will mix several pure substances commonly found in your kitchen with water, vinegar, and a methanol solution containing dissolved iodine and observe the interactions. Then you will use your results to identify several unknown mixtures and to determine the main ingredient in a product that is commonly used to relieve an upset stomach.

Prelaboratory Assignment

- ✓ Read the **Introduction** and **Procedure** before you begin.
- ✓ Answer the Prelaboratory Questions.
  1. What is the difference between a physical and a chemical property?
  2. List three clues used to tell whether a chemical reaction has occurred.
- ✓ If you are not using a Report Sheet for this experiment make a Data Table with the following headings:

Solid	Color	Reaction with water	Reaction with vinegar	Reaction with iodine solution

Materials

Apparatus
Safety goggles
Lab apron
3 thin-stemmed Beral pipets
Masking tape
Marking pen
8 test tubes (13 x 100)
Spatulas
24-well test plate

Reagents
1% iodine solution in methanol
Vinegar
Salt
Baking soda
Baking powder
Sugar (sucrose)
Starch
Black paper

Safety

None of the chemicals (known to chemists as *reagents*) is hazardous, but common-sense procedures should be followed:

1. Wear safety goggles and an apron at all times in the laboratory.
2. Wipe up all spills with large amounts of water.
3. Alcohol solutions can be flammable.
4. Wash your hands thoroughly before you leave the laboratory.

Procedure

1. Label five test tubes: 1, 2, 3, 4, and 5. Label three more test tubes: Mix 1, Mix 2, and Mix 3.

2. Label the three Beral pipets, water, vinegar, and iodine (alc) solution.

3. From the stock bottles provided by your teacher, place a small amount (about the size of a large pea) of each pure substance into the labeled (1 – 5) test tubes that you just prepared. Be sure you put solid #1 in test tube #1, etc. Record the color of each substance in a data table.

4. Place your 24-well test plate flat on the lab bench so that the long edge is vertical toward you. Place a small amount of pure substance 1 in the first three wells of the first row (across). Repeat this procedure with three samples of pure substance 2 in the second row. Continue with this process until you have placed three samples of each substance to be tested in the wells in the same order as shown in the data table.

5. Carefully place the plate on a sheet of black paper.

6. Note the color of each sample, recording your observations in your data table.

7. Draw the correct liquid into each Beral pipet from the containers supplied by your teacher. Carefully place these with their tips up in an empty beaker to hold them until you are ready to use the liquids.

8. Add four drops of water to each substance in the first *column* (vertical) of your test plate.

9. Note any changes that occur and record your observations on your data table.

10. Repeat steps 8 and 9, using the vinegar and alcohol solutions for columns 2 and 3. When you are finished, rinse out your well plate and shake it to dry it .

11. Repeat steps 4 – 10, using samples of the mixtures obtained in your labeled test tubes from your teacher.

Cleaning Up

1. Rinse the well plate, shake it dry and put it away.
2. Wash the test tubes and return them to their proper location.
3. Wash your hands thoroughly with soap and water before leaving the lab.

Analysis and Conclusions

Complete the **Analysis and Conclusions** section for this experiment, either on your Report Sheet, or in your lab report, as directed by your teacher.

1. Compare your data for the pure substances with the data for each of the two solid mixtures. Which of the pure substances shows behavior that is similar to what you saw with the two-solid mixtures (Mixture 1 and Mixture 2)? If you cannot positively identify your mixture, discuss the possibilities for the mixtures.

2. Compare your results for the unknown mixture (Mixture 3) to the observed characteristics of each pure substance. Can you identify any of the pure substances as a component of one of the unknown mixtures? Explain.

3. Based on your answer to question 1, what is the identity for each of your two-solid mixtures?

4. What is the major component of the unknown (Mixture 3), commonly used as an upset stomach- and pain reliever? Compare your results to the ingredients on the package.

5. Two of the pure substances (baking powder and baking soda) are used to make baked products. What chemical property makes these substances useful in baking? How is this property useful in baking? Which two of the five pure substances exhibit this property?

6. Baking soda reacts with acidic solutions, but not with water. Identify which of the five pure compounds is baking soda, and which is baking powder. Explain.

7. One of your pure substances produced a characteristic color (blue-black) when it reacted with the iodine-alcohol solution. Starch is an organic compound you have probably encountered in your biology course. Identify which of the pure substances is starch.

Something Extra

What property do the upset-stomach reliever and the baking ingredients have in common? Give a hypothesis for how this property could possibly produce relief for an upset stomach.

Experiment 6

Properties of Matter

Problem
How can we organize and classify matter so as to better understand the world around us?

Introduction
One of the principle techniques used by chemists is the classification of various types of matter according to their physical and chemical properties. This experiment appears short, but there is a lot to it. You will be given several samples of matter and you are to gather as much information as you can about each one. You will use standard laboratory apparatus: balances, test tubes, and so on, as well as a special device, called a *conductivity tester*. As its name implies, the conductivity tester is used to determine whether or not a solid or liquid can conduct an electric current. Its use will be demonstrated in class.

Prelaboratory Assignment
✓ Read the **Introduction** and **Procedure** before you begin.
✓ Answer the Prelaboratory Questions.
   1. No food or drink should ever be brought into or consumed in the laboratory. Why not?
   2. As the procedure describes, you should never test the odor of a reagent by holding it directly under your nose. Why not?

If you are not using a Report Sheet for this experiment prepare a Data Table with the following headings:

Substance Tested	Physical State	Odor/Appearance	Flammable? (Y/N)	Other Notes

Materials

Apparatus
Lab apron
Safety goggles
Tooth picks
Conductivity tester
Tweezers
Bunsen burner
Soda-straw scoops
Matches
Watch glass
Small test tubes

Reagents
Solid samples: silicon
                    sulfur
                    steel wool (iron)
                    magnesium pieces
                    ammonium chloride
                    sucrose (table sugar)
                    washing soda (sodium carbonate)
                    iodine crystals
                    naphthalene (moth crystals)
                    copper sulfate crystals
Liquid samples: (In dropper bottles)
                    Paint thinner
                    Ethyl alcohol ("grain alcohol")

	Distilled water
	"White" vinegar
	Household ammonia
Testing reagents:	(In dropper bottles)
	Dilute hydrochloric acid (an acid)
	Dilute sodium hydroxide (a base)
	Universal indicator solution (pipet)

Safety

1. Most of the reagents are nontoxic, but common-sense precautions should always be taken when handling chemicals.
2. Due to toxicity, especially of vapors, your teacher will demonstrate the properties of iodine and the combustion of sulfur.
3. Even dilute acids are harmful to skin and clothing. The same is true of bases.
4. Do not taste anything used in the laboratory.
5. Use proper technique when checking for odors: Gently wave the palm of your hand above the surface of the sample to waft vapors toward your nose.

Procedure

There is no specific sequence of steps to be followed, but be sure to investigate the individual properties of each substance separately before investigating how the samples react with each other or with the various testing reagents.

In each stage of the experiment, work with the smallest amount you can. This is especially true with tests of solubility. In many cases, only a small amount of a substance will dissolve in a given amount of liquid – particularly when adding solids to liquids. The more solid you add, the harder it is to tell whether or not a small amount dissolved.

To test the flammability of a solid:	Use tweezers to hold a very small piece in the burner flame.
To test the flammability of a liquid:	Place a few drops on a watch glass and touch the liquid with a lighted match or splint.
Do Not Throw Used Matches In The Sink!	When they have cooled, throw them into a waste basket.

Part 1   Individual properties

For each solid: Describe the appearance (color, texture, etc.) and odor (if any), and determine whether the material is flexible or brittle, a conductor or an insulator, flammable or inflammable. Test to see whether each will react with dilute acids (hydrochloric acid, HCl) or dilute bases (sodium hydroxide, NaOH).

For each liquid: Describe the appearance (color, clarity) and odor (if any). Determine whether the liquid conducts an electric current well, weakly, or not at all. Test the flammability of the liquid as described above in the Materials section. Test about 1 mL of the liquid with one drop of Universal Indicator and record the color. Use the color chart provided with the Universal Indicator to help you decide if the liquid is acidic, basic, or neutral.

Part 2   Solubilities

Note: Many substances dissolve in water, but do so slowly. Stirring helps, but sometimes it takes time. Be patient; don't expect immediate results. On the other hand, if nothing happens within a couple of minutes, the chances are nothing will happen after that.

Solids: Place a very small sample (2-3 crystals, in most cases) on the clean watch glass. Add about 10-15 drops of distilled water and stir. If the sample dissolves, test the solution with the conductivity tester, then add a drop of universal indicator.

For any solid that *does not* dissolve in water, repeat the test using a fresh sample, but instead of distilled water add about 10-15 drops of paint thinner. If the solid dissolves in the paint thinner, test the solution with the conductivity tester and with Universal Indicator.

Liquids: Place about 1 cm of water in a small tube. Add about 4-5 drops (just enough to see) of the liquid to be tested. If it seems to disappear into the water, it's soluble. If it can be seen as a bubble floating on the top of the water (or at the bottom of the tube) it is not soluble.

Part 3   Controls
In separate tubes, test about a milliliter each of sodium hydroxide solution and hydrochloric acid with a drop of Universal Indicator. Observe and record the results.

Cleaning Up
1. **Solid Residues:**
    a. There are separate, labeled containers for iodine and naphthalene.
    b. All other solids should be placed in a waste basket, including cooled matches.
2. **Liquids:**
    a. Paint thinner and mixtures containing paint thinner go in the container labeled, "Organic Waste."
    b. Solutions containing iodine go in the container for iodine waste.
    c. All other liquids can be rinsed down the sink. Acids and bases should be made neutral (as indicated by Universal Indicator solution) before disposal.
3. All glassware and other equipment should be cleaned and returned to their proper locations.
4. Wash your hands thoroughly before leaving the laboratory.

Analysis and Conclusions
Complete the **Analysis and Conclusions** section for this experiment either on your Report Sheet or in your lab report as directed by your teacher.

   Summarize your findings, including (but not limited to) the following suggestions.
   ✓ **Categorize** the samples into groups that show similarities. Clearly identify the reasons behind your groupings
   ✓ What **generalizations** can you make about: solubilities? ability to conduct electric currents? flammability?
   ✓ Discuss any **further tests** that might have given you more information about the various samples.

Experiment 7

Separation Challenge

Problem

Design a procedure to separate a small sample of a mixture of solids into its components and recover each of the components in their original states.

Introduction

You will be given a mixture of sawdust, sand, iron filings, and salt. Your challenge is to design a procedure to separate the mixture into its components and recover each component separately.

Prelaboratory Assignment

✓ Read the **Introduction** and **Procedure** before you begin.

✓ Answer the Prelaboratory Questions.

1. List four separation techniques that might be used in this experiment. Explain how each could isolate a component of the mixture.

2. Draw a flow chart to show how you would proceed to isolate each component in your sample.

Materials

4 test tubes (13 x 100)   **You may use these materials to assist you:**
Mixture sample   Funnel
4 small baggies   Heating unit
Safety goggles   Filter paper
Lab apron   Magnet
   Distilled water
   Forceps
   Thermometer

Safety

None of the chemicals is hazardous, but common-sense procedures should be followed:

1. Wear safety goggles at all times in the laboratory.
2. Wipe up all spills with large amounts of water.
3. Confirm the safety of your procedures with your teacher before proceeding.

Procedure

1. Obtain your sample mixture from your teacher.

2. Design a procedure to isolate each component.

3. Check with your teacher regarding the safety of all of your procedures.

4. Place each component you isolate into a small baggie, seal and label it to turn in to your teacher.

Cleaning Up

1. Clean-up your work area when you are finished.
2. Return all 4 test tubes to your teacher.
3. Wash your hands thoroughly before leaving the laboratory.

Analysis and Conclusions

Complete the **Analysis and Conclusions** section for this experiment either on your Report Sheet or in your lab report as directed by your teacher.

1. Your report should include the following components:
   ✓ Statement of purpose.
   ✓ A list of materials used.
   ✓ Your procedure written in a series of numbered steps.
   ✓ A discussion of your sources of error in both your separation and recovery techniques. Explain how you might proceed differently in the future to eliminate these problems.
2. Use a piece of cardboard ($8^1/_2$ in x 11 in) to create a flow chart for your procedure. Attach the baggies containing your separated samples to the proper places on the chart. Label the chart with your name and turn it in to your teacher.
3. **The physical properties of the substances to be separated are the essential key to a successful separation. Separation techniques are based on these physical properties.** Explain how these statements relate to your challenge. Give several specific examples from your procedure of the property and the separation achieved.

Experiment 8

The Sludge Test

Problem
How can the physical properties of substances be used to separate a mixture into its components?

Introduction
You will be given a mixture containing gravel, sand, and salt. Your goal is to separate the substances and determine the percent of each in the original mixture. You will need to employ the techniques of dissolving, filtration, and evaporation to successfully separate the substances.

Filtration is a process that separates a substance that dissolves from those not dissolved. Evaporation separates a dissolved substance from the substance in which it is dissolved.

Prelaboratory Assignment
✓ Read the **Introduction** and **Procedure** before you begin.
✓ Answer the Prelaboratory Questions.
  1. Explain how a filtration can separate a substance that dissolves from one that does not dissolve. In this experiment what does filtration separate?
  2. What techniques are needed to isolate the salt in this experiment?

Materials
Lab apron	Safety goggles
Balance	250-mL beakers (2)
Filter paper	Heating unit (burner and ring stand, or hot plate)
Pipestem triangle	Watch glass
Small iron ring	Ring stand
Stirring rod	Funnel
Sludge	Forceps

Safety
None of the chemicals (known to chemists as *reagents*) is hazardous, but common-sense procedures should be followed:
  1. Use proper care with open flames in the laboratory.
  2. Wipe up all spills with large amounts of water.
  3. Safety goggles and a lab apron must be worn at all times when working in the laboratory.

Procedure

Part 1   Making Sludge

1. Obtain approximately 10 grams of your sludge resource from your teacher.

2. Carefully determine the mass of your sample. Record the results in your data table.

3. Place the sludge resource in a 250-mL beaker and add about 100 mL of water. Stir the mixture thoroughly to make your sludge for analysis.

Part 2   Separation of Dissolved Sludge Components

1. Determine the mass of a piece of filter paper and initial it for identification purposes.

2. Fold the filter paper into a cone as shown in **Figure 1** and place it into a funnel (opened with 3/4 to one side and 1/4 to the other).

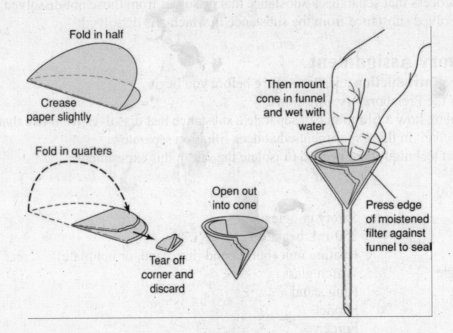

Figure 1
Preparing the filter paper

3. Set up a ring stand with a ring and a clay triangle as shown in **Figure 2**.

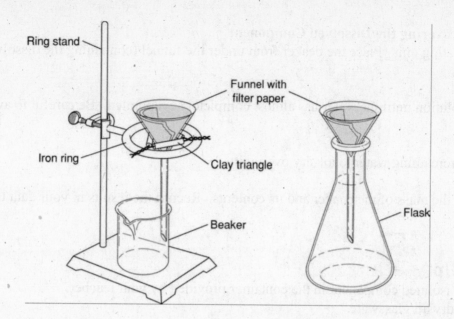

Ring stand

Funnel with
filter paper

Iron ring

Clay triangle

Beaker

Flask

Figure 2
Complete filtration apparatus (two possible set-ups are shown)

4. Find the mass of a second clean, dry, 250-mL beaker, and place it under the funnel.

5. Carefully **decant** the liquid from your sludge into the funnel.
   a. To **decant** a liquid you will need to place the stirring rod across the top of the beaker as your teacher demonstrated. Allow the liquid to pour into the funnel while holding back the majority of the undissolved solids. Pour slowly and evenly to get the best separation.
   b. Be careful to keep the mixture inside the filter paper and not between the paper and the funnel.

6. Carefully add the solids to the filter paper, scraping the beaker to remove any last traces. You can also use a little water to remove any stubborn pieces of solid.

7. Remove the filter paper containing the solid residue from the funnel and place the filter paper on a watch glass. Unfold the filter paper so that the material can be easily dried in the air.

8. Use forceps to remove the large pieces of gravel from the mixture. Place them on a weighed clean piece of filter paper. Determine the mass of the gravel. Record your results in your data table.

9. After drying the wet filter paper, determine the mass of the sand. Record your results in your data table.

Part 3   Recovering the Dissolved Component

1. Set up a heating unit. Place the beaker from under the funnel (containing the dissolved salt) on the unit.

2. Heat the solution until the water has almost completely evaporated. Be careful to avoid spattering.

3. Allow the remaining water to air-dry overnight.

4. Determine the mass of the beaker and its contents. Record the results in your data table.

Cleaning Up

1. Place each isolated component in the container provided by your teacher.
2. Wash and dry all glassware.
3. Return all materials to their proper locations in the lab.
4. Wash your hands thoroughly before leaving the laboratory.

Analysis and Conclusions

Complete the **Analysis and Conclusions** section for this experiment either on your Report Sheet or in your lab report as directed by your teacher.

1. Determine the mass of each component in the sludge mixture and record it in a Summary Table.
2. Determine the percent by mass of each component in the mixture. Show a calculation for one of the components. Record the percent for each component in the Summary Table.
3. Would heating the sludge to begin the separation have been an alternative method for separating the mixture? Explain.
4. Why is it important to avoid spattering the salt solution during the evaporation process?
5. What are possible sources for error in your results?
6. Obtain the true composition of the sludge from your teacher. Discuss how well your results match the true composition of the mixture and the results obtained by other groups.

Something Extra

Propose an alternative method for separating the gravel from the sludge mixture.

Experiment 9

Separation of Mixtures

Problem

How can differences in physical properties be used to separate the components of a solid mixture?

Introduction

Mixtures make up most of the substances with which we come into contact every day. The parts of a mixture (called the *components*) can be separated from one another by taking advantage of the differences in their physical properties. In carrying out the experiment, *Properties of Matter (Experiment 6)*, you learned that substances differ in their abilities to dissolve in various solvents. Some are soluble in water, while others are soluble in different solvents, such as paint thinner. Filtration can be used to separate the insoluble component(s) from those that are soluble in a particular solvent. Soluble components can be recovered by evaporation of the *filtrate*, the liquid which passes through the filter.

Some substances evaporate more quickly than others; for example, paint thinner evaporates faster than water. We say that paint thinner is *more volatile* than water. A few substances are attracted to magnets, while most are not. These, and other differences in physical properties, can help us separate the components of a mixture.

In this experiment you will learn three methods for separating the substances in a mixture. Starting with a sample of a solid mixture, your task will be to separate the mixture into three separate components. You will measure the mass of each component that you recover and compare the total mass recovered to the mass of the original sample.

In principle, one might be able to use a magnifying glass and forceps to separate each individual piece of solid into three separate piles. However, that would not be practical; you will employ much more efficient methods.

Prelaboratory Assignment

✓ Read the **Introduction** and **Procedure** before you begin.
✓ Answer the Prelaboratory Questions.
1. Write a one-sentence explanation of what is meant by a *volatile* substance.
2. Describe the correct procedure for testing the odors of unknown substances.
3. What physical property will make it possible for you to collect one of the mixture components by extraction?
4. If you heat a mixture containing one or more volatile components, what is likely to happen?

Materials

Apparatus
Milligram or centigram balance
10-mL graduated cylinder
10-mL Erlenmeyer flask
Boiling stone
Filter paper
Spatula or wooden stick
Hot sand bath or hot plate
Tongs or hot pad (for handling flask)
Hand lens (optional)
Safety goggles
Lab apron

Reagents
Solid mixture (in stoppered reaction tube)
Distilled water

Safety

1. The sand bath or hot plate will be very hot, (>200°C) but will not appear so. Handle with caution.
2. Before heating, a boiling stone will be added to the flask to reduce the "bumping" that can occur when liquids are boiled.
3. Despite the presence of the boiling stone, some spattering may occur during the evaporation. Direct the mouth of the flask away from yourself and others, and keep the area clear.

Procedure

Note: If a sand bath is to be used, check that it has been turned on, and that the controller is set at 40% power. Hot plates can remain in the "off" position until needed.

1. Find the mass of the stoppered reaction tube containing a sample of the solid mixture and record it in a data table. When you examine the sample, how many components can you visibly distinguish? Record all observations.

2. Add 2 mL of distilled water to the tube. Restopper the tube and shake it vigorously. After a few seconds, feel the sides of the tube. Note and record any temperature change.

3. Measure and record the mass of a 10-mL Erlenmeyer flask containing a fresh boiling stone. Carefully transfer as much of the liquid as possible from the tube into the flask, *without losing any of the solid*. Add about 1 mL of water to the solid remaining in the tube, replace the stopper, and shake the tube again. Once again, transfer just the liquid to the Erlenmeyer flask. The process you have just carried out is called *extraction;* the combined liquids that you poured into the flask is called the *extract*. Save this extract for later use.

4. Measure and record the mass of a single piece of filter paper. Fold the filter paper into fourths, then place the filter paper in a funnel, with a small flask or beaker in position as a receiver. Add about 1 mL of distilled water to the tube, then stopper and shake the tube. Quickly pour the contents of the tube into the filter. Use additional rinses, ~0.5 mL each, as needed to transfer all of the insoluble solid to the filter. Label the filter and solid residue and set it aside for later examination and weighing.

5. Clean and dry the reaction tube and stopper; determine and record their combined masses.
   Note: Although they look identical test tubes vary in mass. The same is true for stoppers. Be sure that you find and record the mass of the tube and the stopper that you used.

6. Place your Erlenmeyer flask on the sand bath. Carefully heat the extract to boiling. The boiling stone prevents boil-over or "bumping" (sudden violent boiling). Maintain a steady rate of boiling while you carry out step 7.

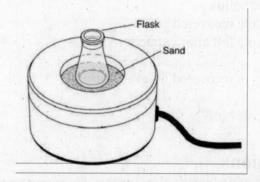

Set-up for Using a Sand Bath to Evaporate a Solution

7. Examine the solid on the filter paper and record your observations. Do all the particles appear to be the same substance? How does their appearance compare with the particles in the original mixture? Your teacher will provide a safe location where the solid can dry overnight or until the next lab period.

8. Continue heating the extract on the hot sand bath until you have evaporated the contents to dryness. No moisture should remain on the walls of the flask at all. If droplets persist, use tongs to lay the flask on the rim of the sand bath. Rotate it occasionally so that all sides are heated, being careful not to get any sand inside the flask. If moisture still persists, ask your teacher for advice.

9. When evaporation is complete, use tongs or a hot pad to remove the flask and cool it for 5-10 minutes. After the flask has cooled to room temperature, brush off any sand from the outside of the flask, and measure, and record the combined mass of the flask and contents.

10. If one is available, use a hand lens to examine the crystals of solid in the flask. How do they compare to the original mixture? Do they look like the solid that was left after the extraction? Record your observations.

11. The following lab day, measure and record the mass of the filter paper and the insoluble residue.

Cleaning Up

1. The solid left in the Erlenmeyer flask may be washed down the drain.
   Note: Be sure to retrieve the boiling stone, which you can place in the waste basket.
2. Return the solid left on the filter paper to your teacher for reuse.
3. Wash all glassware and return it to the proper location.
4. Wash your hands thoroughly before leaving the laboratory.

Summary of Results

Summarize your results in a table that includes:
- ✓ Mass of original sample
- ✓ Mass of insoluble residue
- ✓ Mass of soluble residue
- ✓ Total mass of recovered solids
- ✓ Percent of original sample recovered
- ✓ Percent of original sample left after extraction ("insoluble residue")
- ✓ Percent of original sample recovered by evaporation ("soluble residue")
- ✓ Percent of original sample "lost"

Analysis and Conclusions

Complete the **Analysis and Conclusions** section for this experiment either on your Report Sheet or in your lab report as directed by your teacher.

1. How does the appearance of each of the recovered solids compare with each other and with the original mixture? Be complete. Note that you are being asked to make three comparisons, not just one.

2. **a.** Does the insoluble residue appear to be one substance, or more than one? Explain.
   b. Given your answer to 2a, and recalling that you were told that the original mixture consisted of three components, how many of the three components of the original mixture must have been in the liquid extract? Defend your answer.

3. Consult with at least three other groups. How do their results for percent of material recovered compare with yours?

4. Account for the material "lost" during the experiment. In what part of the procedure was it most likely lost? (**Hint:** Recall the discussion of "volatile" substances in the introduction and in the prelaboratory questions.)

Something Extra

The answers to these questions are based on the results obtained by the class as a whole.

1. Enter your values for the percent (by mass) of the insoluble residue, and the soluble residue in the Class Data Table, as directed by your teacher. Also enter the percent of original sample that was "lost" during the experiment.

2. Calculate the mean (average) values for the percentages of soluble and insoluble residue, as well as the mean value for percentage lost.

3. Calculate the average percent deviations for each of the three fractions.

4. Examine the three percentages and their deviations.
   a. What is the sum of the three percentages? Ideally, what should the sum be? Why?
   b. Suggest one or more reasons why the sum of the percentages might be different from what you predicted in **4a**. Based on your reasons, would the sum of the individual percentages be more likely to be greater than 100% or less than 100%. Explain.
   c. By how much might you expect the sum of the percentages to differ from 100%? By 1%? By 5%? By 10%? Discuss.
   d. Discuss the precision of this experiment and make suggestions as to where the procedure might be changed so that all of the groups would report similar results.

Experiment 10

Distillation

Problem

How does a difference in physical properties make it possible to separate the components of a solution?

Introduction

Chapter Two of *World of Chemistry* discusses matter and the various physical and chemical changes it can undergo. The chapter also points out that pure substances can be separated from each other on the basis of the differences in their physical and chemical properties.

Distillation is cited as an example of a physical process that can be used to separate two miscible liquids from each other. (*Miscible* means that the two liquids mix uniformly with each other; they form a solution.) The separation occurs because each of the liquids boils at a different temperature. Liquids with low boiling points are said to be "*more volatile*" while those with high boiling points are called "*less volatile*."

In a simple distillation, the mixture is heated until the solution boils, giving off the vapor of the component with the lower boiling point (the more volatile of the two). The vapor is directed through a tube cooled by air or water (known as a *condenser*) into another container (called the **receiver**). Gentle heating is continued until all (or most) of the more volatile liquid has evaporated and been condensed in the receiver. The pure liquid collected in this fashion is called the *distillate*. If there are only two liquids in the mixture, distillation is complete at this point. In more complex situations, such as the separation of crude oil into the various petroleum fractions, a more elaborate separation and collection process is necessary. *Since distillation involves no change in the identity of either component, it is a physical process, rather than a chemical one.*

You will carry out a distillation, then do some tests of the physical properties of the distillate to help you confirm the validity of the sentence in italics.

Prelaboratory Assignment

✓ Read the **Introduction** and **Procedure** before you begin.
✓ Answer the Prelaboratory Questions.
   1. Why is it important that the distilling flask not be allowed to go to dryness?
   2. Why is it important to place the tip of the temperature probe well below the sidearm of the distilling flask? (**Hint:** What temperature is it supposed to be measuring?)
   3. What is the function of the ice water in the beaker that holds the receiver vial?
   4. Would distillation be a satisfactory means of separating a liquid from a dissolved solid? Explain.

Materials

Apparatus

Side-arm distilling flask
Liebig condenser (or similar)
Hot plate
Heat-transfer block (optional)
Boiling stone
Graduated cylinder, 25-mL
Watch glass
Beaker, 30-mL or similar
CBL2, LabPro, or similar interface
temperature probe
Graphing calculator, TI-83+, or similar
Safety goggles
Lab apron

Reagents

Distillation mixture
Ice water or crushed ice
Cobalt chloride

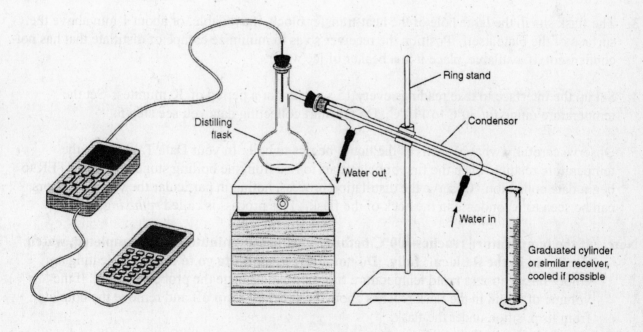

Distilling
flask

Ring stand

Condensor

Water out

Water in

Graduated cylinder
or similar receiver,
cooled if possible

Figure 1
Distillation apparatus

Safety

1. **Keep open flames away from the distilling apparatus.** One of the liquids in the mixture is extremely flammable. In case of fire, call your teacher. **Do not** try to put the fire out yourself.
2. The hot plate can get very hot; do not touch it.
3. Use great care when inserting the side arm of the distilling flask into the rubber stopper connecting the flask to the condenser.

Procedure

1. Place a boiling chip in the 125-mL distilling flask and add 40 mL of the water-methanol mixture to the flask. Note the odor and appearance of the mixture and record your observations in a Data Table.

2. Attach the flask to the condenser using the rubber stopper on the side-arm. Insert the stopper with the temperature probe into the top of the flask. Be sure that the stopper is as high as it can be on the probe, so that the tip of the probe extends as far below the side-arm as possible. This ensures the best possible contact of the distilling vapors with the sensor.

3. The flask sits in the large hole of the heat-transfer block, if available, or about 1 mm above the surface of the plate itself. Position the receiver so as to minimize escape of distillate that has not condensed. If available, place it in a beaker of ice water.

4. Set up the interface to take readings every 15 seconds for a period of 30 minutes. Set the temperature range for 30°C to 110°C. Do not start collecting data yet; see step 5.

5. Observe carefully what happens as the liquid begins to boil. In your Data Table, note the temperature reading when the first bubbles start to rise from the boiling stone. Press ENTER to begin data collection. Observe the distillation process, noting in particular the way the vapors can be seen to recondense in the neck of the flask. This process is called *refluxing*.

Note: **If the temperature reaches 100°C before the entire 30-minute run is completed, watch the liquid in the flask carefully. Do not allow the flask to go to dryness.** Heating an empty flask causes a rapid temperature rise than can damage the probe severely. If the volume of liquid in the flask reaches about 5 mL or less, turn off and remove the hotplate from its position under the flask.

6. When distillation is complete, turn off the hotplate and carefully remove the receiver. Note and record the volume of distillate recovered. If the receiver has been sitting in an ice-water bath, dry the outside thoroughly, then determine the mass of the graduate and distillate.

Disconnect the flask from the condenser and reposition the flask well above the hot-plate so it can cool. Transfer about 4-5 mL of your distillate to the small beaker. Pour the remaining liquid into the container labeled Methanol Distillate. Dry the graduate and determine its mass alone.

In your Data Table, enter the volume of methanol recovered and the mass of the graduated cylinder, with and without the methanol.

7. Place **four or five drops** of the distillate on a Pyrex watch glass and set it on fire. Do the same with a few drops of the original mixture of water and methanol. Be sure the distilling flask has cooled sufficiently to be handled, then repeat the flammability test with a few drops of the residue left in the flask.

   In each case, record whether all, part, or none of the sample is consumed. Compare the results with the mixture to those you got for the pure distillate and the residue. What differences (if any) do you find? How do you explain any differences?

   Record your data for all three liquids in your Data Table.

8. Place 10-15 drops of the distillate in a small tube and add 1-2 small crystals of cobalt(II) chloride, $CoCl_2$. Observe the color the solid gives to the liquid. Repeat the cobalt chloride test with small amounts of the original mixture and of the residue in the flask. Record the test results in Data Table 2.

Cleaning Up
1. **After everything has cooled enough to handle safely**, place the distillate and any liquid left in the distilling flask in the container provided so that it can be recycled.
2. The boiling chip is to be placed in the waste basket, **not** in the sink.
3. If your teacher directs you to do so, disassemble the apparatus and put the pieces away.
4. Wash your hands thoroughly before leaving the laboratory.

Analysis and Conclusions
Complete the **Analysis and Conclusions** section for this experiment either on your Report Sheet or in your lab report as directed by your teacher.

1. Use Graph Link or a similar program to transfer the graph of temperature vs. time to a computer. Print the graph and attach it to your report.

2. Use the TRACE button on your calculator to trace along the graph as it appears on the screen. On the graph, indicate the following:
   a. the temperature at which you began recording
   b. the time at which the graph began its steep rise
   c. the time and temperature when the graph began to level
   d. the temperature range during the time when most of the distillate was recovered

3. Provide a *written* description of the shape of your graph.

4. a. How did the flammability of the distillate compare with that of the residue left in the distilling flask and of the original mixture of methanol and water? Is it possible to distinguish among the three liquids on the basis of flammability? Explain.
   b. Summarize your findings when each of the three liquid fractions was tested with cobalt(II) chloride. Is it possible to distinguish among the three using the results of these tests? Explain.

5. During distillation, the temperature of the escaping liquid remains relatively constant while most of the liquid distills. This temperature is the boiling point of the distillate.
   a. At what temperature did most of the methanol distill in your experiment?
   b. How does this temperature compare with the accepted boiling point of pure methanol, which is 65°C?
   c. What can you conclude about the identity of the liquid left in the flask?

Something Extra

1. In addition to the flammability and cobalt chloride tests, density can also help in identification of an unknown liquid. Use data from your experiment to determine the density of your distillate. The accepted value for the density of methanol is 0.791 g/mL. Determine the percent deviation between your experimental value for the density of the distillate and the accepted value. (Divide the deviation between the values by the accepted value. Convert the resulting decimal fraction to a percent.)

2. Originally, you set up the CBL to use stored calibration data. Discuss the value of carrying out a manual calibration, rather than using stored data. How might you conduct such a calibration? In what sort of experimental situations might it be preferable to manually calibrate the CBL instead of relying on stored data?

Experiment 11

Aluminum Atoms

Problem
How thick is a piece of aluminum foil? How many atoms are stacked up to make that thickness of the foil?

Introduction
Aluminum is an element we use in the form of aluminum foil in everyday life. We want to find out just how many aluminum atoms need to be stacked up to make a piece of aluminum foil. We will assume that the aluminum atoms are stacked on top of each other directly, and that the atoms behave as solid spheres during the stacking process.

Prelaboratory Assignment
✓ Read the **Introduction** and **Procedure** before you begin.
✓ Answer the Prelaboratory Question.
   1. What measurements will you need in order to determine the thickness of a piece of aluminum foil?

Materials
Lab apron	Balance, milligram or centigram sensitivity
Safety goggles	Metric ruler
Scissors	Aluminum foil
Graduated cylinder	Aluminum block

Safety
1. Wear safety goggles and a lab apron at all times in the laboratory.
2. Wipe up all spills.

Procedure
1. Determine the mass of the aluminum block.

2. Fill the graduated cylinder about halfway with water and record the volume of water accurately.

3. Tilt the cylinder to about a 45 degree angle (be careful not to spill any water), and carefully slide the aluminum block into the graduated cylinder. Set the cylinder upright and be sure that the block is completely covered with water. Record the new volume.

4. Remove the block from the cylinder and dry it with a paper towel, pour out the water and return both the block and the cylinder to your teacher.

5. Obtain a piece of aluminum foil, and cut it if necessary to obtain a square that is approximately 15 cm x 15 cm.

6. Measure the length and width of the piece of foil exactly and record it on your data table.

7. Determine the mass of the piece of aluminum foil and record it on your data table.

Cleaning Up

1. Clean up your station and return materials to their proper places. Ask your teacher about re-using or recycling the aluminum foil.
2. Wash your hands before leaving the laboratory.

Analysis and Conclusions

Complete the **Analysis and Conclusions** section for this experiment either on your Report Sheet or in your lab report as directed by your teacher.

1. Calculate the volume of the aluminum block from the apparent change in the volume of the water in the cylinder.

2. Since both the aluminum block and the aluminum foil are pure elemental aluminum, we would expect the ratio of the mass to the volume to be the same for both. That is:

$$\frac{\text{mass of block}}{\text{volume of block}} = \frac{\text{mass of foil}}{\text{volume of foil}}$$

Use this relationship to find the volume of the aluminum foil.

3. Calculate the thickness of the aluminum foil (**Hint:** Think about how you would calculate the volume of a box from its measurement. Think of the piece of aluminum foil as a very thin box.)

4. One aluminum atom has a diameter of 0.000000025 cm. How many atoms thick is the aluminum foil?

5. What are the possible sources for error in your experiment?

Something Extra

Look up the diameters for lithium, sodium, potassium, and cesium atoms. What is the relationship between the atomic number of the element and the diameter of its atoms?

Experiment 12

Electrolysis

Problem
How can we break a compound down into the elements from which it is formed? How do the properties of the compound compare with those of the individual elements?

Introduction
Chapter 2 in *World of Chemistry* discusses matter and the various physical and chemical changes it can undergo. The chapter goes on to discuss the fact that pure substances can be separated from each other on the basis of the differences in their physical and chemical properties.

In contrast to distillation (discussed in Chapter 2) *electrolysis* is used to break a compound down into the basic elements from which it was formed. The separation (which is really a decomposition), is carried out by passing an electric current through the compound to be broken down. *In this case, unlike distillation, the identity of the original compound does change, so electrolysis is a chemical process.*

In order to speed up the electrolysis, by improving the current flow through the solution, sodium sulfate, Na_2SO_4, has been added to the distilled water. The water also contains a small amount of bromothymol blue. Bromothymol blue is called an *indicator,* because it changes color, depending on the pH of the system. Bromothymol blue is yellow in acid and blue in base. Following completion of your part of the experiment, your teacher will demonstrate the same process on a larger scale.

Prelaboratory Assignment
✓ Read the **Introduction** and **Procedure** before you begin.
✓ Answer the Prelaboratory Questions.
   1. Suggest an explanation for the green indicator color before the electrolysis begins. Remember that distilled water and sodium sulfate are neither acidic nor basic (see the **Introduction**).
   2. Write the formula for water. Write the formulas for the new substances formed in this experiment. Be sure to include the physical states for all substances.

Materials
Apparatus
U-tube
9-volt battery
Battery cap or wire leads and pencil-lead electrodes
Safety goggles
Lab apron

Reagents
Distilled water (with Na_2SO_4 and bromothymol blue)

Safety

1. While there is little danger of personal injury, due to the low level of electrical energy being used, to protect the life of the battery, you should be careful not to touch the two electrodes to each other.
2. Safety goggles and a lab apron must be worn at all times in the laboratory.

Procedure

1. Fill the tube with the water/Na_2SO_4/bromothymol blue mixture to within about 1 cm of the top. If wire leads and pencil-lead electrodes are to be used, insert the electrodes in the tube and connect the leads to the battery as shown in Figure 1. If a battery cap is used, and if the wires from the cap do not have alligator clips on them, insert the ends of the wires directly into the ends of the U-tube.

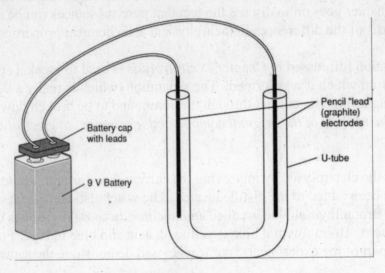

Battery cap with leads

9 V Battery

Pencil "lead" (graphite) electrodes

U-tube

Figure 1
Electrolysis apparatus

2. Allow the current to flow through the solution in the tube for about 5 minutes, or until you are certain that no further changes will occur. You are looking for any evidence that a chemical change is taking place. Record your observations.

Cleaning Up

1. Disconnect the wires or battery clip from the battery. Return battery and wires to the proper location.
2. The solution in the tube is harmless. It can be rinsed down the drain with water.
3. Clean the U-tube thoroughly with water (not just a quick rinse) and return it to its original location.
4. Clean the pencil leads (if used) by wiping with paper towel. If one breaks, dispose of it in the container for non-hazardous solid waste.
5. Wash your hands thoroughly before leaving the laboratory.

Analysis and Conclusions

Complete the **Analysis and Conclusions** section for this experiment either on your Report Sheet or in your lab report as directed by your teacher.

1. There are ample clues for chemical (as well as physical) changes in the process which you have just carried out. Cite at least three indicators of chemical change that you observed in this experiment.

2. Following the completion of your portion of the experiment, your teacher performed two demonstrations which illustrated important characteristics of the products of the decomposition.
   a. What does the 'rocket launch' tell you about a mixture of hydrogen and oxygen gases?
   b. In the second demonstration, for which the plastic pipet bulb was clamped in place, why did the bulb fill with water? What difference between gases and liquids does this illustrate?

Experiment 13

Classifying Elements
An Introduction To The Periodic Table

Problem
What characteristics can be used to divide the chemical elements into groups with similar properties?

Introduction
The one hundred or so chemical elements can be roughly divided on the basis of their physical and chemical behavior. For example, many of the elements are *brittle*, meaning that they will break, rather than bend, when subjected to pressure or stress. Elements that will bend under stress are described as *malleable*.

Another means of categorizing an element is by physical appearance: Is it colored? Does the surface have a shiny luster or does it have a dull finish? Still, a third classification is the element's ability to conduct an electric current.

All three of these represent *physical properties* since they deal with the way the element itself behaves on its own, in the absence of other substances.

Chemical properties, on the other hand, have to do with the way in which one element or compound interacts with other pure substances. The reaction with acids or other chemicals are examples of tests of the chemical properties of an element.

In this experiment, you will run five different tests on solid samples of eight separate elements, each of which will be identified only by code letters, **a-h**. Three of the tests will involve physical properties, while the other two will illustrate chemical behavior. Tests will be carried out in the wells of a 24-well plastic test plate, except as otherwise indicated. Element **h** will be distributed by your teacher, since it is unsafe to handle directly. The other seven elements are quite safe, although reasonable precautions (such as washing your hands thoroughly before leaving the laboratory) should always be taken.

Your teacher will also demonstrate the use of a conductivity tester to measure the ability of each of the elements to conduct an electric current.

The physical properties will be studied first because the reaction with other chemicals may change the physical properties of the element being tested. As you recall, chemical properties involve chemical changes, which change the identities of some or all of the reactants.

Due to difficulties in handling element **h**, your teacher will demonstrate the tests for brittleness and for conductivity during the pre-lab discussion. You will do the other tests on **h** yourself.

Prelaboratory Assignment

✓ Read the **Introduction** and **Procedure** before you begin.
✓ Answer the Prelaboratory Questions.

1. Consult the periodic table on the inside back cover of your text. Based on the elements you are familiar with, where are most of the metals found?
2. Look at the numbering system for the groups as explained in Chapter 3 of your textbook. Do the "A" groups contain mostly metals or mostly nonmetals? Do the groups in the middle of the table, (transition elements) contain mostly metals or mostly nonmetals?
3. What basis did Mendeleev use to divide the elements into *groups*?

✓ If you are not using a Report Sheet for this experiment construct a Data Table for Part 1 with the following headings:

Element	Malleable or Brittle?	Conducts?	Shiny or Dull?	Color or Other Characteristics

Materials

Apparatus
24-well test plate
Forceps or tweezers
Conductivity tester
Distilled water wash bottle
Safety goggles
Lab apron

Reagents
0.5 M HCl(*aq*)
0.5 M CuCl$_2$(*aq*)
Samples of elements **a-h**

Safety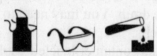

1. Handle hydrochloric acid with care; it is corrosive to skin and clothing.
2. At least one of the solid element samples is toxic, and will stain skin and clothing. Avoid touching the samples; wash your hands thoroughly with soap and water before leaving the laboratory.
3. Safety goggles and a lab apron must be worn at all times in the laboratory.

Procedure

Part 1   Tests of Physical Properties

1. Place small pieces of each of the elements **a** through **g** in separate wells of your test plate using forceps or a small spatula. Record the color and appearance; is the surface dull or shiny? (In other words, does it have a luster?)

2. Using tweezers or forceps, try bending each sample. Will it bend (is it malleable), or does it snap or crumble into pieces, showing brittleness? If you cannot get it to bend or break, remove the sample from the well, set it on a clean, hard surface (like a piece of paper on the base of a ring stand) and try hitting it gently with an iron ring or some other instrument, such as a small hammer. Does pounding cause the sample to flatten (malleability) or does the solid shatter (brittleness)? Record your results in your Data Table.

3. Test the electrical conductivity of your sample by touching the two wire ends of your conductivity tester to each sample. If your tester uses only a single LED, decide if it lights brightly, weakly, or not at all. If you are using a tester with multiple LEDs note how many (if any) of them light up. If your tester uses sound, record the loudness of the sound. Record your results in the Data Table.

Part 2 Tests of Chemical Properties.

4. Divide your samples so that you have two wells containing element **a**, two with element **b**, and so on. You may need to get more of some of the elements, but these tests can be done with very small samples; you need only be able to see whether changes are taking place.

5. To one of the wells containing element **a,** add just enough dilute hydrochloric acid, HCl(*aq*), to cover the solid; you do not need to fill the well. Look for any evidence of chemical reaction, such as color changes (in either the sample or the acid), evolution of a gas, or even significant changes in temperature, although those will be hard to detect with such small quantities.

6. Test the other sample of **a** with copper(II) chloride solution, CuCl₂(*aq*). Again, look for any sign of chemical change. In the same fashion, test the rest of the solids, **b** through **h**, with hydrochloric acid and with copper(II) chloride solution. Record all of your results in your Data Table.

Cleaning Up

1. While most of the solids are not dangerous, they are not water soluble, so you must keep the pieces from falling into the sink where they can wash down the drain. You may need to use tweezers or forceps to remove solid pieces from some of the wells.
2. Use a wash bottle to rinse your test plate onto paper towels in the container labeled "Solid Reagent Waste." Wash the test plate thoroughly with soap and water and return it to its place. If one or more wells were stained during the experiment today, your teacher may be able to show you how to clean it. For there to be any chance to remove stains, the cleaning must be done before the stain has a chance to set into the plastic, so don't wait to clean up your test plate.
3. Wash your hands thoroughly before leaving the laboratory.

Analysis and Conclusions

Complete the **Analysis and Conclusions** section for this experiment either on your Report Sheet or in your lab report as directed by your teacher.

1. For each of the five tests of physical and chemical properties, group the eight elements according to their behavior on that particular test.
   ✓ Malleable or brittle?
   ✓ Shiny or dull?
   ✓ Conducts?
   ✓ Reacts with HCl(*aq*)?
   ✓ Reacts with CuCl₂?

2. Combine your five lists from Question 1 into two groups of elements such that all the members of a given group are alike in at least most of the properties tested. You may find that you have one or two elements that don't clearly belong to either of your two categories because some properties fit one group, while other properties fit the second group better. If that happens, make them a third category.

3. Describe briefly the criteria you used to make your groupings. Identify any difficulties you encountered in deciding where to place each element.

4. Using the definitions for metals, nonmetals and metalloids from your textbook, and your groupings, try to identify each of the elements **a** through **h** as being a *metal*, a *nonmetal*, or a *metalloid*.

Experiment 14

Electric Solutions

Problem
Which compounds contain ions?

Introduction
Some compounds contain ions. We can test for this by dissolving a compound in water. If the compound contains ions, the ions will be dispersed throughout the water and are free to move. The movement of the ions in a solution enables them to conduct a current.

In this lab you will be using a conductivity tester to determine which compounds contain ions.

Prelaboratory Assignment
✓ Read the entire experiment before you begin.
✓ Answer the Prelaboratory Questions.
   1. Why does the bulb light? What does it mean if the bulb doesn't light?
   2. Why did we add nothing to a beaker of water and test it?

Materials

Apparatus	Reagents
Safety goggles	HCl (0.10 *M*)
Lab apron	Potassium nitrate
Conductivity tester	Sodium chloride
Beakers (250 mL) or cups (8 oz), (11)	Vinegar
Plastic spoon	Sugar
	Water

Safety
1. Safety goggles and a lab apron must be worn at all times in the laboratory.
2. If you come in contact with any solution, wash the contacted area thoroughly.

Procedure
1. Half-fill six beakers or cups with water.

2. Add the following to each cup
   Cup 1:   nothing
   Cup 2:   spoonful of sugar
   Cup 3:   spoonful of sodium chloride
   Cup 4:   spoonful of potassium nitrate
   Cup 5:   10 mL of 0.10 *M* HCl
   Cup 6:   10 mL vinegar

3. Use the conductivity testers provided by your teacher to test each cup. Carefully place the tips of the tester into the solution. Record your observations.

4. For the solutions that caused the bulb to light, pour a small amount (1 or 2 mL) of each solution into separate cups. Add water to each cup until it is almost filled.

5. Use the conductivity testers to test each of your diluted solutions. Record your observations.

Cleaning Up
1. Clean up all materials and wash your hands thoroughly.
2. Dispose of all chemicals as instructed by your teacher.

Data/Observations
Complete the Data/Observations section for this experiment either on your Report Sheet or in your lab report as directed by your teacher.

1. In which cases did the bulb light? If the bulb lit, was it rather bright or dim?
2. What happens when water is added to the solutions that originally caused the bulb to light?

Analysis and Conclusions
Complete the **Analysis and Conclusions** section for this experiment either on your Report Sheet or in your lab report as directed by your teacher.

1. How are the compounds similar that caused the bulb to light?

2. True or false: The reason a compound did not cause the bulb to light is because the substance did not dissolve in water. Explain your answer.

3. What do your observations tell you about the contents of each cup? Draw molecular level pictures for each cup to explain your results.

4. Explain what happened when you added water to the solutions that originally caused the bulb to light by using molecular level pictures.

5. How could we tell if a compound consisted of ions if it does not dissolve in water?

Something Extra
Test several household products to see if they contain ions. Get permission from your teacher first.

Experiment 15

Forming and Naming Ionic Compounds

Problem
When do ionic substances react to form a product? What are the names and formulas for these products?

Introduction
When they dissolve in water, ionic compounds break apart into ions. These ions move about among the water molecules bumping into the other ions and molecules in the solution. When two ionic solutions are mixed several things can happen! In this experiment you will have an opportunity to mix various ionic compounds in solution. When solutions of some ionic compounds are mixed, the cation from one and the anion from another form an insoluble compound which appears as cloudy or grainy solid, called a *precipitate*. On the other hand, if all cation-anion combinations form soluble pairs, no precipitate appears. All of the ions remain in solution.

In this experiment your task will be to mix ions of different kinds and to observe whether they form precipitates. If a precipitate is formed you will write the formula for the new compound and then name the product.

Prelaboratory Assignment
✓ Read the **Introduction** and **Procedure** before you begin.
✓ Answer the Prelaboratory Questions.
  1. For the following pairs of ions, write the formula of the compound that you would expect them to form:
     a. barium and hydroxide
     b. cobalt(III) and phosphate
     c. iron(II) and sulfate
     d. silver and hydrogen carbonate
  2. Platinum is a transition metal and forms Pt^{2+} and Pt^{4+} ions. Write the formulas for the compounds for each of these ions with:
     a. bromide ions
     b. carbonate ions.

Materials

Apparatus
Lab apron
Safety goggles
96-well tray or acetate sheet
Toothpicks
Paper grid

Reagent solutions
Sodium phosphate
Potassium hydroxide
Sodium oxalate
Cobalt chloride
Strontium chloride

Reagent solutions

Potassium ferrocyanide	Nickel(II) chloride
Sodium carbonate	Potassium iodide
Silver nitrate	Lead(II) nitrate
Copper(II) sulfate	Iron (III) nitrate

Safety

1. Wear safety goggles and lab aprons at all times in the laboratory.
2. No food or drink is allowed in the laboratory at any time.

Procedure

1. Obtain a well tray or an acetate sheet and a paper grid. If you are using the acetate sheet place a paper grid on the table and cover it with the acetate sheet.

2. Obtain a set of the solutions to be tested. These may be in bottles or pipets. It may be helpful to organize these according to their positions on your paper grid.

3. In each box or well place three to five drops of each solution corresponding to that box. Record on your Data Table or a copy of the reaction grid any reaction that occurs or cloudiness that develops. Note both the color and texture of the precipitate (use a toothpick to help determine texture). In those cases where no reaction occurs, mark NR(no reaction) on your data sheet.

4. Continue mixing and recording until your entire data table is complete.

Cleaning Up

1. Clean the accumulated precipitates and solutions directly from the test plate or plastic sheet into the container indicated by your teacher.
2. Thoroughly rinse the test plate or sponge off the plastic sheet.
3. Dispose of all used toothpicks in a waste basket.
4. Wash your hands thoroughly before leaving the laboratory.

Analysis and Conclusions

Complete the **Analysis and Conclusions** section for this experiment either on your Report Sheet or in your lab report as directed by your teacher.

1. For each case in which you found a reaction occurred, write the correct formula for the substance formed.

2. For each formula write the correct chemical name for the compound formed.

Experiment 16

Grocery Store Nomenclature

Problem
Can you determine which compounds are in your consumer products?

Introduction
Chemicals are not just for chemistry class. We use chemicals everyday and in everything we buy. In this activity you will determine the identity of a chemical based on clues given to you by your teacher. You will then find out which products contain these chemicals.

Prelaboratory Assignment
✓ Read the entire experiment before you begin.

Materials
List of clues
References

Procedure
1. Obtain a list of clues to the identities of 5 chemical compounds.

2. Use references (including your text, other books, and the Internet) to help you determine the identity of each chemical.

3. Find products in which each of the chemicals is present. You may find these at home but you may also need to go to a grocery store or drugstore.

Summary Table
List the name, formula, and products for each of the compounds. Include justification for each. Report this information either on your Report Sheet or in your lab report as directed by your teacher.

Something Extra
Contact a company that produces a product that contains one of the chemicals on your list and find out the purpose of the chemical in the product.

Experiment 17

Energy Changes In
Physical and Chemical Systems

Problem

How should we interpret the energy changes that accompany physical and chemical processes?

Introduction

In this brief investigation you will make observations of two processes. The first involves dissolving a solid in water, while the second is the reaction between baking soda and vinegar. So that you can detect the temperature changes that accompany the two processes, both will be carried out in a small, thin-walled plastic cylinder or zip-lock plastic bag.

Your primary focus in this experiment is on the energy changes that occur. You will monitor those changes simply by noting whether the container feels cold or hot. If the system is releasing heat, the container will feel warm to the touch. On the other hand, if it feels cold to your touch, the container and contents must be drawing heat from your body.

Prelaboratory Assignment

✓ Read the **Introduction** and **Procedure** before you begin.
✓ Answer the Prelaboratory Questions.
   1. What distinguishes a physical change from a chemical change?
   2. When something feels cold to the touch, in what direction is energy being transferred to or from your body?

Materials

Apparatus
Safety goggles
Plastic cylinder or zip-lock bag
Small scoop
Plastic stirring stick
Wash bottle with distilled water
Lab apron

Reagents
Calcium chloride pellets, $CaCl_2$
Sodium bicarbonate, $NaHCO_3$ (baking soda)
Vinegar

Safety

None of the chemicals (known to chemists as *reagents*) is hazardous, but common-sense procedures should be followed:
1. Wear safety goggles and a lab apron at all times in the laboratory.
2. Wipe up all spills with large amounts of water.

Procedure

Part 1   Physical Changes

1. Place two small scoops of calcium chloride, $CaCl_2$, in the clear plastic graduated cylinder or bag.

2. Add distilled water to the 5 mL mark. Stir with the plastic stirring stick.

3. Describe what happens, including any temperature changes, evidence of reaction, etc, in your notebook or on your Report Sheet.

4. Clean the cylinder by rinsing the contents down the drain. Be sure all traces of the first reaction are removed from the cylinder, then rinse it with distilled water. You need not dry the inside.

Part 2   Chemical Changes

5. Place two small scoops of baking soda, sodium bicarbonate, $NaHCO_3$, in the clean cylinder.

6. Vinegar is a water solution of a weak acid, known as acetic acid, $HC_2H_3O_2$. Use a dropping bottle to add vinegar, a little at a time, until the liquid level reaches about 5 mL.

7. Describe what happened, including any temperature changes, evidence of reaction, etc., in your notebook or on your Report Sheet.

Cleaning Up

1. Clean the plastic cylinder as you did in step 4.
2. Rinse and dry the scoop, make sure the caps are back on the containers and return everything to its original location.
3. Wash your hands thoroughly before leaving the laboratory.

Analysis and Conclusions

Complete the **Analysis and Conclusions** section for this experiment either on your Report Sheet or in your lab report as directed by your teacher.

Part 1   Calcium Chloride and Water
1. a. When you touched the cylinder after adding water to the calcium chloride, was energy being released or absorbed by the contents? Explain.
   b. Was energy being released or absorbed by your hand? Explain.

Part 2 . Vinegar and Baking Soda
1. a. What evidence do you have that something new is being formed in this case?
   b. When you touched the cylinder, was energy being released or absorbed by the contents. Explain.
   c. Was energy being released or absorbed by your hand? Explain.

2. Any process that releases energy is said to be *exothermic*. Any process that takes in (absorbs energy) is described as being *endothermic*.
   a. When calcium chloride, $CaCl_2$ dissolves in water is the process exothermic or endothermic? Explain.
   b. Is the reaction between baking soda, (sodium bicarbonate, $NaHCO_3$), and vinegar, (dilute acetic acid, $HC_2H_3O_2$), exothermic or endothermic? Explain.

Experiment 18

Measurement and the SI System

Problem

What is the best way to record and use the uncertainty inherent in laboratory measurements?

Introduction

Experimental observations often include measurements of volume, mass, length and temperature. There are important characteristics for any measurement: (1) *Accuracy* – how close the value is to the correct value and (2) *Precision* – the agreement of several measurements of the same type, which is a reflection of the degree of uncertainty involved in the measurement.

In this experiment you will be focusing primarily on the uncertainty of measurements. The uncertainty of a measurement is often represented by the symbol, "±" (read as "plus or minus") followed by an amount. Thus, if an object is measured as being 25 cm long, and the length is known to within 0.01 cm, its length might be reported as:

$$25.00 \text{ cm} \pm 0.01 \text{ cm}.$$

With measuring instruments such as rulers or graduated cylinders, readings are generally estimated to one more decimal place than the smallest scale division. Thus, a ruler that is graduated in millimeters allows you to estimate to the nearest 0.1 mm.

Measurements in which the uncertainty is a result of a visual estimate are called *analog* measurements. Electronic devices, such as balances and thermometers, give a *digital* reading. In such cases, the last decimal place reported represents the uncertainty. Thus, if a balance gives a mass reading of 3.004 g, you know that the uncertainty is ± 0.001 g, or ± 1 mg.

Sometimes a piece of equipment will have scale markings that are only approximate and are not intended to be used for precise work. This is true of the beakers and Erlenmeyer flasks that you will use in this course. In such cases, neither the accuracy nor the precision of the measurements is high. The container will usually carry a label to this effect.

Although the activities in this experiment are numbered, they can be done in any order. Your teacher may assign different groups to begin with different parts of the Procedure. If not, organize your work to adjust for availability of equipment

Prelaboratory Assignment

✓ Read the **Introduction** and **Procedure** before you begin.
✓ Answer the Prelaboratory Questions.
1. If you are using a graduated cylinder for which the smallest division is 0.1 mL, to what degree of precision should you report liquid volumes? Express your answer in the form:

$$\pm \underline{\quad \text{(uncertainty)} \quad} \text{ mL}$$

2. A stack of five 3.5-inch floppy disks is 1.6 cm tall.
   a. What is the average thickness of one disk?
   b. To the nearest whole number, how many disks will be in a stack 10 cm tall?
3. In Celsius degrees:
   a. what is the boiling point of water?
   b. what is the freezing point of water?
   c. what is normal human body temperature?
4. What is a meniscus; what role does it play in the correct reading of liquid volumes?

Materials

Apparatus
Centigram or milligram balance
Metric ruler
50-mL beaker
100-mL beaker
25-mL graduated cylinder
100-mL graduated cylinder
Thin-stem pipet
Plastic weighing cups
Thermometer, -20°C to 110°C
Safety goggles
Lab apron

Reagents
Tap water
Pennies, post 1983 and pre 1982
Nickel
Ice, crushed or small pieces
Rock salt

Safety

1. Wear safety goggles and lab aprons at all times in the laboratory.
2. No food or drink is allowed in the laboratory at any time.
3. If you are using a mercury thermometer, be very careful. Mercury is extremely toxic. If you break a mercury thermometer, report it to your teacher immediately.

Procedure

Part 1   Reading Volumes
There are six graduated cylinders, each labeled and each containing some quantity of liquid to which food coloring has been added to make the volume easier to read. Record the capacity of each cylinder and the volume of liquid it contains in a Data Table. Remember to include units (mL).

In your Data Table, record the uncertainty involved in each volume measurement. Remember that the uncertainty depends on the size of the divisions; it is always one-tenth of the smallest division.

Part 2   Comparing Measuring Containers

Use tap water to fill a 50-mL beaker to the 20-mL mark. Use a thin-stem pipet to adjust the level until the bottom of the meniscus is lined up with the 20-mL line.

Pour the water from the beaker into a 25-mL graduated cylinder. Read the volume according to the 25-mL graduate and record it in a second Data Table. Finally, transfer the water from the 25-mL graduate to a 100-mL graduate. Read the volume again and record it in your Data Table.

Part 3   Measuring Coins

Use a metric ruler to determine the diameter of a penny and a nickel. Report the diameters in both centimeters (cm) and in millimeters (mm). To the nearest whole number, determine the number of pennies it takes to make a stack that is 1 cm tall. Find the mass of this number of pennies. Record these measurements in a third Data Table.

Part 4   Comparing Pennies

Determine the masses of two pennies. Make sure that one is dated before 1982, and the other is 1983 or later. Record the masses in a fourth Data Table.

Part 5   Effect of Salt on the Temperature of an Ice-Water Mixture

Half-fill a 100-mL beaker with crushed ice. Place a thermometer in the melting ice and hold it there until the temperature is no longer changing (usually about 15-20 seconds). At this point, the thermometer and ice/water are said to be in "thermal equilibrium." Read and record the temperature in a fifth Data Table. Add one level plastic spoonful of rock salt. Do not stir the solution. When the temperature is again constant, read and record the final temperature of the water-ice-salt mixture.

Cleaning Up

1. There is nothing except water and salt to dispose of, and these can safely go down the drain
2. Rinse the beaker and thermometer. Return all equipment – including glassware, coins, and rulers – to their proper locations.
3. Be sure to place any broken glassware in the container specifically labeled for this purpose.
4. Wash your hands thoroughly before you leave the laboratory.

Analysis and Conclusions

Complete the **Analysis and Conclusions** section for this experiment either on your Report Sheet or in your lab report as directed by your teacher.

1. It is common to get different volume readings from each container in Part 2. What explanation can you offer for:
   a. an apparent decrease in volume?
   b. an apparent increase in volume?

2. Which container in Part 2 gave you the most *precise* reading of the actual volume of water it held? Justify your choice. (**Hint:** Remember the distinction between precision and accuracy given in the Introduction.)

3. The beaker you used in Part 2 probably carries the notation "± 5%." What do you interpret this to mean?

4. Calculate the average thickness of a single penny using the data you obtained in Part 3 of the procedure. Show your calculations.

5. Explain how you could estimate the mass of a large stack containing an unknown number of pennies using only the data from Part 3.

6. Determine the number of pennies that could be laid edge-to-edge the full length of a meter stick. Repeat the calculation for nickels. (Hint: Have you ever seen a third of a penny? Do fractional coins exist?)

7. Compare your results from Part 4 with those of other teams. How do the masses of pennies minted before 1982 compare with the masses of newer ones? Try to explain the difference.

8. Errors or variations from expected results that *do not* result from carelessness or incorrect procedure are called *random experimental errors*. Random experimental errors are no one's fault; they are unavoidable and they must be taken into account any time we evaluate the results of an experiment. Suggest two sources of random experimental errors that might cause different teams to get different results in Part 4 of the procedure.

9. What effect does salt have on the temperature of an ice-water mixture? Did other groups observe the same effect?

Experiment 19

Measuring A Book? Precisely!

Problem
How does the precision of an instrument affect measurements and calculations?

Introduction
Measurement is an important part of science, and an understanding of uncertainty is an important part of measurement.

In this activity you will compare measurements of your textbook with four different rulers, each with a different precision.

Prelaboratory Assignment
✓ Read the entire experiment before you begin.

Materials
4 special rulers
Chemistry textbook

Procedure
1. Measure the length and width of your chemistry textbook with each of the four special rulers. Record these values.

2. Convert all measurements to centimeters (if necessary). Record these values.

3. Using your measurements, calculate the perimeter (in cm) and area (in cm²) of the cover of your chemistry textbook. Record these values.

Cleaning Up
Leave the rulers on the lab bench.

Analysis and Conclusions
Complete the **Analysis and Conclusions** section for this experiment either on your Report Sheet or in your lab report as directed by your teacher.

1. Which ruler gives you the most precise measure of the perimeter and area of the cover of your chemistry textbook? Why?

2. Which ruler gives you the least precise measure of the perimeter and area of the cover of your chemistry textbook? Why?

3. Justify the number of significant figures in each of your measurements.

4. Justify the number of significant figures in each of your calculations.

5. Compare your measurements with other groups. For which ruler was there the most difference among groups? The least difference? Explain.

6. Compare your calculations with other groups. For which ruler was there the most difference among groups? The least difference? Explain.

Something Extra
For each ruler, what is the largest length that you can measure that you would report as a measurement of zero? Explain.

Experiment 20

Conversion Factors

Problem
How can measured values be converted from one type of unit to another?

Introduction
In this experiment you will be introduced to the derivation and use of *conversion factors*. As the name implies, conversion factors are a means to change from one unit of measure to another. Some of these factors are familiar from everyday use: there are seven days in one week, twelve inches equals one foot, and so on. In each case, the same quantity is represented in two different ways. While seven days is the same length of time as one week, no one would suggest that the numbers themselves, seven and one, are equal – that would be absurd! Thus,

$$7 \text{ days} = 1 \text{ week} \quad \text{but} \quad 7 \neq 1$$
$$12 \text{ inches} = 1 \text{ foot} \quad \text{but} \quad 12 \neq 1$$
$$100 \text{ centimeters} = 1 \text{ meter} \quad \text{but} \quad 100 \neq 1$$

In each example you find the same quantity represented by two different measurements. But *are* they different? No, and it is **this distinction between numbers and measurements** that is the focus of this experiment. You are to include the units in all reported measurements; without the units, all you have is a number – no physical quantity has been represented.

Graphs are a powerful tool for showing the relationship between measured quantities. As part of your Analysis and Conclusions, you will graph the data you have collected. The graph will then be used to determine values that were not included in your Observations and Data.

Prelaboratory Assignment
✓ Read the **Introduction** and **Procedure** before you begin.
✓ Answer the Prelaboratory Questions.

1. Describe the procedure for determining the slope of a straight-line graph.
2. Explain the difference between a *measurement* and a *number*.
3. Every conversion factor can be written in two ways. Complete the table below.

7 days	= 1 week	1 day = 0.143 week ($1 \div 7 = 0.143$)
12 inches	= 1 foot	1 inch = _____ foot
100 cm	= 1 m	1 cm = _____ m
$1	= 10 dimes	1 dime = $_____

Materials

Apparatus
Milligram or centigram balance
Thin-stem pipet
30-mL beaker (for holding water)
10-mL graduated cylinder
Lab apron
Safety goggles

Reagents
Distilled water

Safety

1. Wear safety goggles and aprons at all times in the laboratory.
2. No food or drink is allowed in the laboratory at any time.

Procedure

1. Find the mass of a clean, dry 10-mL graduated cylinder to the nearest milligram. If the only cylinder available is wet inside, you should dry it completely before you begin. Record this mass and all subsequent measurements in a Data Table.

2. **Remove the graduated cylinder from the balance,** then use a beaker of distilled water and a thin-stem pipet to add precisely 1.50 mL of water to your graduated cylinder. The low point of the curved liquid surface (the *meniscus*) should just line up with the 1.50-mL mark on the scale. Read at eye level to be certain that you gauge the volume correctly. Hold the pipet vertically so that the drops will fall freely to the bottom of the cylinder. If drops are left on the side walls of the graduate, your results will be inaccurate.

3. Find the mass of the cylinder containing the 1.50 mL of water, and record the total mass in your Data Table.

4. Again remove the graduated cylinder from the balance then fill the graduate to exactly 3.00 mL and find the total mass once again. Repeat for 5.00 mL.

5. Use the thin-stem pipet to add 20 more drops of water to the contents of the cylinder. Estimate the volume to the nearest 0.01 mL, and record it in the data table. (Do not determine the mass of the cylinder and contents.)

6. Finally, add an additional 20 drops to the cylinder and read the volume once again, estimating as before to the nearest 0.01 mL.

Cleaning Up

1. Pour the water down the sink.
2. Return the glassware to its proper location.
3. Dispose of the pipet as your teacher directs.
4. Wash your hands thoroughly before leaving the laboratory.

Analysis and Conclusions

Complete the **Analysis and Conclusions** section for this experiment either on your Report Sheet or in your lab report as directed by your teacher.

1. Use the information in your Data Table to calculate the mass of water in 0.00 mL, 1.50 mL, 3.00 mL, and 5.00 mL. Show your calculations and enter the results in a Summary Table.

2. Use the graph paper to plot the data from **1**. Label the axes, with mass on the vertical axis and volume on the horizontal axis. Draw the best-fit straight line through the four data points.

3. Determine each of the following (**a** through **e**). Summarize your results in a second Summary Table.
    a. Graphing your data allows you to find values you didn't actually measure. Use your graph to find **(i)** the mass of 2.40 mL of water and **(ii)** the volume of 4.25 g of water.
    b. Calculate the *slope* of your graph. This is the density of water.
    c. Based on your results for steps 5 and 6 of the procedure, what is the average volume for the number of drops listed: (Drops = 20, 100, 50, 10, 1)?
    d. Use the density you determined for water to find:
    (i) The total mass of water in the graduate after the first 20-drop addition.
    (ii) The total mass of water in the graduate after the second 20-drop addition.
    (iii) The average mass of 20 drops of water?
    e. What is the mass of:
    (i) 1 drop of water?
    (ii) 10^6 drops ?
    (iii) 6.02×10^{23} drops?

4. Use your experimental results to complete the following conversions:
    a. 1 drop of water = _____ g = _____ mL
    b. 1 gram of water = _____ mL = _____ drops
    c. 1 milliliter of water = _____ g = _____ drops

5. If you did not start with a completely dry graduated cylinder in step 1, would this affect your answers? If so, what part of the answer would be incorrect: the mass of water, the volume of water, or both? Explain.

6. If the drops of water from the pipet hit the sides of the graduate, how could this affect your answers? Again, be specific in describing what can go wrong.

7. The density of water is often assumed to be exactly one gram per milliliter. How close to that value is the slope of your graph? If your value was not within 5% of the accepted value, that is, within the range of 0.95 g/mL ≤ density ≤ 1.05 g/mL, was there anything about the procedure that made it difficult for you to get a precise result? Explain.

Something Extra

1. The values obtained by the class for the density of water will be listed on the board. Use the posted information to calculate the mean value for water's density, the average deviation and percent deviation.

   a. The mean is simply the arithmetic average of all the individual results. Show your calculations.

   b. Deviation refers to the difference between the mean and each individual value. Deviations are reported as absolute values, so do not include signs. Show two sample calculations of deviations, one for a value greater than the mean and one for a value less than the mean.

   c. As the name implies, the average deviation is the arithmetic mean of the individual deviations. Show your calculations.

   d. Combine your answers to **1a** and **1c** to express the experimental value for the density of water in the form (mean) ± (average deviation). Round both portions of the answer to the appropriate decimal place.

2. Use your graphing calculator to determine both the mean value for the density of water and the standard deviation from that mean. Report the density of water in the form: (mean) ± (standard deviation). Round both portions of the answer to the appropriate number of decimal places.

Experiment 21

Measurement and Density

Problem

How can measurements be used to determine the density of a variety of regular and irregular objects? How do different methods of making measurements compare?

Introduction

Careful measurements are required for quantitative scientific observation. Chemists use the metric system of units for making measurements in the laboratory. Making measurements requires estimations because the limitations of the equipment necessitate estimating the final digit in the measurement.

Prelaboratory Assignment

✓ Read the **Introduction** and **Procedure** before you begin.
✓ Answer the Prelaboratory Questions.
  1. Describe how to determine the limitation of a metric ruler.
  2. Why is it important to read the graduated cylinder at eye level on the bench top?
  3. Why is it important to be sure that no drops of water cling to the cylinder in the water displacement method?

Materials

Wooden or metal blocks	100-mL graduated cylinder
Metric ruler	250-mL beaker
Irregular shape solid	Knife
New red potato	
Safety goggles	
Lab apron	

Safety

1. Wear safety goggles and lab aprons at all times in the laboratory.
2. No food or drink is allowed in the laboratory at any time.
3. Be careful when using the knife. It is sharp.
3. Wash your hands thoroughly before leaving the laboratory.

Procedure

Part 1   Length

  1. Select any three blocks.

  2. Use the metric ruler to measure the length, width and height of each block in centimeters. Record your measurements to the limitations of the metric ruler.

Part 2   Volume

3. Pour tap water into the 100-mL graduated cylinder up to the 85 mL mark. Notice the curve in the center of the water level. This curve is called the *meniscus*. The volume of a liquid is read from the bottom of the meniscus. To obtain a proper reading your eye level should be the same as the water level. Be sure to place the graduated cylinder on a flat surface before reading the meniscus. Move your eye to the level of the water in the cylinder, do not lift the cylinder.

4. **Measuring the volume of a beaker.**
   a. **Method 1 – Water Capacity**
      1. Fill the 250-mL beaker to the very top.
      2. Empty the beaker in a series of steps by using the 100-mL graduated cylinder, keeping track of the total amount of water. Record the total volume of water from the beaker.

   b. **Method 2 - Formula**
      1. Measure the height of the beaker.
      2. Measure the diameter of the beaker.
      3. Consider the beaker to be a cylinder. Using the formula for the volume of a cylinder ($V = \pi r^2 h$, learned in mathematics), calculate the volume of the beaker.

5. **Volume of a regular shaped solid - Formula**
   Use your measurements from Part 1 to calculate the volume of each of the blocks. Use the formula for a rectangular solid ($V = l \times w \times h$).

6. **Volume of an irregular shaped solid – Water Displacement Method**
   a. Fill your 100-mL graduated cylinder between ½ and ¾ full. Record the volume of the water exactly.
   b. Carefully place the irregular object into the water. Tilt the cylinder and slide the object into the water slowly. Be sure the object is completely submersed and that there are no drops clinging to the side of the cylinder. Record the new water level.
   c. The volume of the object is the difference between the two water levels. Record the volume of the irregular object.

Part 3   Mass

7. There are many different types of balances used in the chemistry laboratory. Your teacher will demonstrate how to use the balances in your classroom to determine the mass of an object.

8. Use the balance to find the mass of each block you measured in Part 1.

9. Use the balance to determine the mass of your irregular shaped object.

Part 4   Density

10. Obtain a new red potato from your teacher.

11. Use the skills you have learned in the lab and your knowledge of density to determine the density of the potato.

Experiment 23

Decomposing Copper Oxide

Problem
How is copper oxide formed from copper metal? What is the formula of the copper oxide used in this experiment?

Introduction
Metals react with oxygen in the air to form metallic oxides. The process of rusting, for example, involves the reaction of iron and oxygen. As you learned in Chapter 4, many metals can form more than one type of cation. For example, copper (Cu) can form Cu^+ ions or Cu^{2+} ions in ionic compounds. You will be given a sample of copper oxide and will determine whether it is copper(I) oxide or copper(II) oxide.

In this lab you will heat copper oxide powder in the absence of oxygen to form copper metal. From the masses of the original copper oxide and the copper metal you obtain, you can calculate the percent by mass of copper in the original compound. By comparing your result to the percent by mass of copper in copper (I) oxide and copper (II) oxide, you can decide which compound was the starting material.

Prelaboratory Assignment
✓ Read the entire experiment before you begin.
✓ Answer the Prelaboratory Questions.
   1. Calculate the percent copper by mass in copper(I) oxide.
   2. Calculate the percent copper by mass in copper(II) oxide.

Materials
Apparatus
Safety goggles
Lab apron
Test tube (13 x 100 mm)
Bunsen burner
Matches
Balance
Ring stand
Test tube clamp for ring stand
2-holed rubber stopper fitted with glass tubing
Rubber tubing

Reagents
Copper oxide powder

Safety

1. Safety goggles and a lab apron must be worn at all times in the laboratory.
2. You will work with a flame in this lab. Tie back hair and loose clothing.
3. Do not drop matches into the sink. Dispose of burned matches in the trashcan after they are cool.

Procedure

1. Accurately determine the mass of your test tube.

2. Measure between 3.00 and 4.00 g of copper oxide powder and place it in the test tube.

3. Clamp the test tube to the ring stand so that the tube is slightly angled. Make sure there is a thin coating of copper oxide on the test tube and that the copper oxide is away from the mouth of the test tube.

4. Assemble the apparatus as shown in the **Figure 1.** The gas should flow through the test tube and into the Bunsen burner.

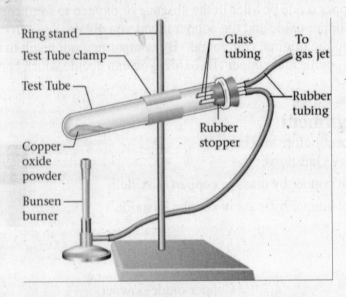

Figure 1
Apparatus for Decomposing Copper Oxide

5. Light the Bunsen burner.

6. Use the hottest part of the flame to heat the copper oxide in the test tube. Move the Bunsen burner back and forth over the length of the test tube containing the copper oxide. Make sure to keep the flame away from the rubber stopper.

7. Continue heating until all the copper oxide has reacted.

8. Allow the test tube and contents to cool in the absence of oxygen (move the Bunsen burner from under the test tube, but allow it to continue burning until the tube has cooled completely).

9. Accurately measure the mass of the test tube and copper. Determine the mass of copper formed.

Cleaning Up

1. Clean up all materials and return them to the proper locations.
2. Dispose of the copper as instructed by your teacher.
3. Wash yours hands before leaving the laboratory.

Analysis and Conclusions

Complete the **Analysis and Conclusions** section for this experiment either on your Report Sheet or in your lab report as directed by your teacher.

1. How could you tell when the reaction was completed?

2. Calculate the percent copper by mass in your copper oxide.

3. Compare your result from **2** to the mass percent of copper in copper(I) oxide by finding the difference between the two. **Note:** you calculated the mass percent of copper in copper(I) oxide in the Prelaboratory Questions.

4. Compare your result from **2** to the mass percent of copper in copper(II) oxide by finding the difference between the two. **Note:** you calculated the mass percent of copper in copper(II) oxide in the Prelaboratory Questions.

5. What is the formula of the copper oxide you decomposed? Justify your answer.

6. Which would have a larger mass, an iron bar before or after it rusts? Explain your answer.

Something Extra

Why was the gas sent through the test tube? Could this lab be performed with a test tube open to the air? Explain your answer.

Experiment 24

Formula for a Hydrate

Problem
What is a hydrate? What does the formula of a hydrate show? What is the percentage of water in a hydrated salt?

Introduction
A hydrate is a compound that contains water in its crystal structure. The water may be removed from the salt in the laboratory by heating the salt. The salt without the water is called an anhydrous salt. Here are some examples of hydrates:

$$CaSO_4 \cdot 2H_2O \qquad CoCl_2 \cdot 6H_2O \qquad MgSO_4 \cdot 7H_2O$$

In a hydrate the water molecules are a distinct part of the compound but are joined to the salt by connections that are weaker than the connections in the salt or the connections in the water molecules. Notice we use a dot to connect the water units to the salt formula.

In this experiment you will measure the mass of a hydrated salt, such as barium chloride or copper(II) sulfate, then remove the water from the crystals and measure the mass of the anhydrous salt. The data gathered will allow you to determine the percent water in the hydrated salt and also the empirical formula for the hydrated salt.

Prelaboratory Assignment
✓ Read the **Introduction** and **Procedure** before you begin.
✓ Answer the Prelaboratory Questions.
  1. What information is necessary to determine the percentage of water in your hydrate sample?
  2. How will the water be removed from the hydrate in this experiment?
  3. A hydrate has the formula of $MgSO_4 \cdot 7H_2O$. What is the percent water in this hydrate?

Materials
Apparatus
Heating unit
Porcelain crucible
Clay triangle
Safety goggles
Lab apron
Hot pad

Reagents
Hydrated salt

Safety

1. Hot items look the same as cool items. Be sure to wait until the crucible is cool before transferring it from place to place.
2. Wear laboratory safety goggles and a lab apron at all times in the laboratory.

Procedure

Part 1   Preparation

1. Clean a porcelain crucible and dry it by gently heating it. When it is cool, find the mass of the crucible.

2. Obtain a sample of a hydrated salt and find its mass.

3. Place the salt in the crucible and place the crucible in the clay triangle suspended in the iron ring as shown in **Figure 1**.

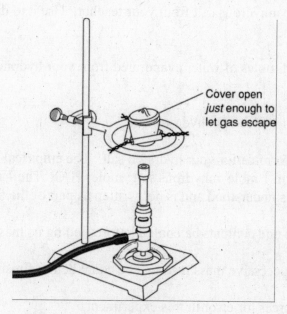

Cover open *just* enough to let gas escape

Figure 1
Apparatus for heating a hydrate

Part 2   Evaporation of Water

1. Heat the crucible, gently at first, and then increase the flame to full intensity. Heat the sample for 10 – 12 minutes.

2. Shut off the flame and allow the crucible to cool completely.

3. Find the mass of the crucible and its contents.

4. Repeat steps 1 – 3 until at least two successive masses are equal within the uncertainty of your balance. This indicates that all of the water has been removed from the crystals.

Cleaning Up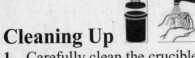

1. Carefully clean the crucible and dry it. Return it to its proper location.
2. Place samples of hydrate in the container provided by your teacher for solid waste.
3. Clean up your lab station, returning all equipment to its proper location.
4. Wash your hands thoroughly before leaving the laboratory.

Analysis and Conclusions

Complete the **Analysis and Conclusions** section for this experiment either on your Report Sheet or in your lab report as directed by your teacher.

1. Determine the percentage of water in the hydrated salt. Think about the fact that the difference between the mass of the hydrated salt and the mass of the anhydrous salt is the mass of the water in the hydrate. Remember the definition of percent.

2. Obtain the name of your anhydrous salt from your teacher. Use it to determine its formula and molar mass.

3. Determine the number of moles of water evaporated from your hydrate. Think about how mass is related to moles.

4. Determine the number of moles of anhydrous salt.

5. Calculate the empirical formula for your hydrated salt. The empirical formula for a hydrate is always written in the form 1 mole anhydrous salt · moles H_2O. The 1 mole in front of the anhydrous salt formula is understood and is not written as part of the formula.

6. Why should the crucible and contents be cooled before finding its mass?

7. Why must at least two successive mass readings be equal before finishing the experiment?

8. List several possible sources for error in this experiment.

Something Extra

How could the water be placed back into the crystals? Propose an experiment to test your hypothesis. After checking with your teacher carry out the experiment and report your results.

Experiment 25

Recognizing Chemical Reactions

Problem
What types of evidence do we have for chemical reactions?

Introduction
Any process in which one or more new substances is formed is called a *chemical change*. Typically, evidence for such a change includes one or more of the following: production of a gas (*effervescence*); formation of a solid product (in microscale experiments often seen as a faint cloudiness in what was a clear solution--this solid is called a *precipitate*); an unexpected color change; or a significant increase or decrease in temperature, as heat is either given off (*exothermic reaction*) or absorbed (*endothermic reaction*).

Your task in this experiment is to make accurate observations of any changes that occur when you mix five chemical *reagents* (re-ay-junts) in pairs. Your ultimate goal is to decide whether those changes signal the occurrence of a chemical reaction.

Prelaboratory Assignment
✓ Read the **Introduction** and **Procedure** before you begin.
✓ Answer the Prelaboratory Questions.

1. What is the distinction between the words "clear" and "colorless" as they apply to solutions? Is a clear solution necessarily colorless? Is a colorless solution necessarily clear? Explain.

2. Explain the difference between chemical changes and physical changes.

3. In complete sentences, define the following terms:
    a. precipitate
    b. reactant
    c. exothermic reaction

Materials
Apparatus
96-well test plate
Tweezers (for handling solid **E**)
Safety goggles
Lab apron

Reagents
Solutions **A, B, C, D**
Solid **E**

Safety

1. Wear safety goggles and a lab apron at all times in the laboratory.
2. Because you don't know the identities of the five reagents, you must assume they are hazardou.
   In fact, this is the case:
   - One is an acid; it is corrosive to skin and clothing.
   - One is mildly toxic, and will stain skin and clothing.
   - One is caustic.

Procedure

1. At your workstation there is a set of four thin-stem pipets, each containing one of the solutions, **A-D**. There is also find a small container of solid **E**. Make observations of these substances ar record your observations in a Data Table.

2. Mix three drops of solution **A** with three drops of solution **B** in one well of your 96-well test plate. Continue in this fashion until you have tested all possible pairs of solutions, recording your observations in a Data Table with the headings: **Substances mixed, Observations, Reaction?** and **Interpretation**. (Because you are concerned with possible reactions between pairs of reactants, you should not mix three or more together at one time.)

3. Not all reactions are rapid. If no change is evident at first, observe again after about 4 minutes to see if gradual changes have taken place. To detect any temperature changes, place your fingertip on the underside of each well before and after mixing.

4. Next, use tweezers to select individual pieces of **E,** placing one piece in each of four wells in your test plate. Add 8-10 drops of a different solution to each piece of **E**. Record your observations in your Data Table.

5. When you have completed steps 1-4, compare your results with those of other groups. If you think further tests are called for, complete those tests before cleaning up and returning to the classroom for the post-lab discussion.

Cleaning Up

1. Use your wash bottle to rinse all liquids into the waste container which has been lined with paper towels.
2. Wash the test plate with soap and water, rinse with tap water and set on paper towels to dry. Make sure that no solid materials go into the sink.
3. Wash your hands thoroughly before leaving the laboratory.

Analysis and Conclusions

Complete the **Analysis and Conclusions** section for this experiment either on your Report Sheet or in your lab report as directed by your teacher.

1. How did your results compare with those of other groups? Suggest possible explanations for any significant differences that you observed.

2. If you were to repeat this experiment, how would you change the procedure? Why would you make that change (or those changes)?

xperiment 26

Examples of Chemical Reactions

oblem

at are some clues that accompany chemical changes?

troduction

emical changes are accompanied by signals, many of which are visual. We can represent emical reactions by writing chemical equations. Because atoms are neither created nor destroyed, must balance these equations.

this lab you will carry out four chemical reactions and make careful observations to determine ich clues accompany chemical changes.

relaboratory Assignment

Read the entire experiment before you begin.
Answer the Prelaboratory Questions.

Write and balance chemical equations for the following reactions. Include the state for each substance in the equation.

a. Solid magnesium reacts with aqueous hydrochloric acid to produce aqueous magnesium chloride and hydrogen gas.
b. Aqueous sodium chloride reacts with aqueous silver nitrate to produce solid silver chloride and aqueous sodium nitrate.
c. Solid magnesium burns in the air to form solid magnesium oxide.

Materials

Apparatus
Safety goggles
Lab apron
Bunsen burner
Matches
Tongs
Wooden splint
Well plate (6 wells)
13 x 100 mm test tubes (2)
100-mL beakers (2)
Spatula

Reagents
Magnesium ribbon
Hydrochloric acid (3.0 M)
Iron (III) nitrate (0.1 M)
Potassium thiocyanate (0.1 M)
Sodium chloride
Silver nitrate

World of Chemistry   77

Safety

1. The 3.0 M HCl is corrosive. Handle it with extreme care.
2. Safety goggles and a lab apron must be worn at all times in the laboratory.
3. Silver nitrate solution stains skin and clothing. Avoid contact with the solution.
4. If you come in contact with any solution, wash the contacted area thoroughly.
5. You will work with a flame in this lab. Tie back hair and loose clothing.
6. Do not drop matches into the sink. Dispose of burned matches in the trashcan after they are c

Procedure

Part 1   Reacting magnesium and hydrochloric acid

1. Place about 5 mL of 3.0 M aqueous hydrochloric acid in a test tube.

2. Add a 2-3 cm piece of magnesium ribbon to the acid. Make and record careful observations.

3. Place another test tube over the mouth of the test tube containing the acid and magnesium (the mouths should be the same size).

4. After the reaction is finished, light a wooden splint. Quickly turn over the top test tube (so the mouth is pointing up) and place the burning wood splint near the mouth. Do not point the test tube toward anyone. Make and record careful observations.

Part 2   Reacting sodium chloride and silver nitrate

1. Dissolve a spatula tip amount of each of the solids in a small amount of water (about 10 mL separate 100-mL beakers.

2. Place a few drops of aqueous sodium chloride into one of the reaction wells.

3. Slowly add a few drops of aqueous silver nitrate to the aqueous sodium chloride. Make and record careful observations.

Part 3   Reacting magnesium and oxygen

1. Light the Bunsen burner.

2. Use tongs to hold a 1-2 cm piece of magnesium ribbon in the flame. Do not look directly at the magnesium as it burns. Make and record careful observations.

Part 4 $Fe^{3+} + SCN^- \rightarrow FeSCN^{2+}$

1. Place aqueous iron (III) nitrate into a clean reaction well. Use enough to cover the bottom of the well.

2. Slowly add aqueous potassium thiocyanate, one drop at a time, to the aqueous iron (III) nitrate until a change is observed. Make and record careful observations.

Cleaning Up

1. Clean up all materials and return them to the proper locations.
2. Dispose of all chemicals as instructed by your teacher.
3. Wash your hands thoroughly before leaving the laboratory.

Analysis and Conclusions

Complete the **Analysis and Conclusions** section for this experiment either on your Report Sheet or in your lab report as directed by your teacher.

1. For each of the reactions, which observations indicate that a chemical reaction was taking place?

2. What are some clues that accompany chemical changes?

Something Extra

List three chemical reactions from everyday life that demonstrate the clues that you observed in this lab.

Experiment 27

Interpreting Chemical Reactions
The Reaction between Copper and Nitric Acid

Problem

What is the best way to describe and represent what takes place during a chemical reaction?

Introduction

Just exactly what takes place during a chemical reaction? For example, when an Alka-Seltzer tablet is dropped into a glass of water, there's a lot of fizzing and some solid sinks to the bottom of the glass, but there is less solid remaining than before. Was this a chemical reaction? How can you tell? Where did the rest of the solid go? What was the gas? Where did it come from? When you strike a match, a flame appears and seems to consume the match. Where did the match go? We could ask the same questions about the combustion of a gas like propane, or a liquid like gasoline. In these cases, *everything* seems to disappear. Is it just gone, or is it hiding somewhere?

This activity will help you to understand better what is happening in chemical reactions. Your task is to make careful observations of the changes that occur when nitric acid reacts with a small piece of copper wire and to record those observations. By relating your observations to the equation for the reaction, you will get a better understanding of what happens in a chemical reaction.

As part of the prelaboratory discussion, your teacher will clean a length of copper wire by rubbing it with fine steel wool to rid the surface of any corrosion that may have tarnished the wire surface. The wire will then be cut into short lengths, 3-5 mm each, for your use in the experiment.

Prelaboratory Assignment

✓ Read the **Introduction** and **Procedure** before you begin.
✓ Answer the Prelaboratory Questions.

1. Use the words, *reactants* and *products* in a sentence which shows that you understand the meanings of the terms.

2. The procedure tells you that a gas will be produced as part of the reaction. Should you expect to see bubbles coming out of the delivery system? Explain.

Materials

Apparatus
Test tube (13 x 100 mm)
Thin-stem pipet
50- or 100-mL beaker, 3/4 full of water
Parafilm™, strip 1-2 cm by 7-8 cm
Safety goggles
Lab apron

Reagents
Copper wire, 0.3-0.5 cm
7 *M* nitric acid, about 0.5 mL

Safety

1. Nitric acid is extremely corrosive to skin and clothing. Handle it with extreme care
2. The gas generated during the reaction is highly toxic. Avoid breathing it. **At no time should the tube be opened, except in the fume hood, once the reaction has begun.**
3. Your teacher will take care of disposal. Do not flush the contents down the sink.
4. Safety goggles and a lab apron must be worn at all times in the laboratory.

Procedure

1. Use a pair of scissors to cut the bulb of the pipet as shown below. Start in the middle of the bulb and cut at about a 45° angle away from the stem. The top of the bulb and the stem will be placed over the mouth of the test tube to provide a sealed delivery tube to safely remove a toxic gas that is produced during the reaction.

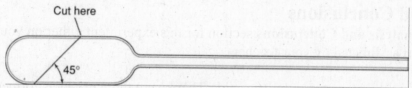

Figure 1
Cutting a thin stem pipet bulb

2. Place one small piece of the wire in the 13 x 100-mm test tube.

3. Take the tube, delivery system, and a 100-mL beaker of water to the fume hood, where your teacher will add about 0.5 mL of 7 M nitric acid, HNO_3.

4. Immediately fit the delivery system over the mouth of the test tube and bend the stem into a 180° arc as shown in **Figure 2**. Wrap the strip of Parafilm™ around the bottom of the pipet bulb to help seal in the gas that is produced. Put the stem into the beaker of water. The gas that is produced, NO_2, is quite toxic, but is very soluble in water. This system prevents it from escaping into the air. Carry the entire system back to your work station.

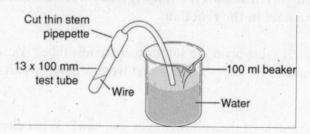

Figure 2
Completed Assembly

5. Observe the reaction that occurs, making note of all the evidence you can find that proves that what you are seeing is indeed a chemical reaction. Hold a piece of white paper behind the tube; what do you see? Be sure to keep the delivery system under the level of the water during the entire time you observe the reaction.

6. While you are waiting for the reaction to finish, begin to work on the **Analysis and Conclusions** questions.

7. If time permits, you can return to the lab near the end of the period to see if any further changes have occurred.

Cleaning Up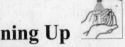

1. After a few minutes, when you are certain that no further changes are going to occur, return the tube to the fume hood. You will find a 24-well test plate in the fume hood. Stand the test tube in the 24-well plate, noting the letter and number that identify the well in which you place your tube.

2. Wash your hands thoroughly with soap and water before you leave the laboratory.

Analysis and Conclusions

Complete the **Analysis and Conclusions** section for this experiment either on your Report Sheet or in your lab report as directed by your teacher

The reactants were copper metal and nitric acid. There were three products formed by the reaction: an aqueous solution of copper(II) nitrate, nitrogen dioxide gas, and water. All five of the participating species are represented below, first in a word equation, then as a molecular equation; neither of the equations is balanced.

| Copper metal | + | Nitric acid solution | → | Copper(II) nitrate solution | + | Nitrogen dioxide gas | + | Water |

$$Cu(s) \quad + \quad HNO_3(aq) \quad \rightarrow \quad Cu(NO_3)_2(aq) \quad + \quad NO_2(g) \quad + \quad H_2O(l)$$

In each of the questions that follow, identify all substances by their name <u>and</u> by their symbol or formula, including their physical states: (*s*), (*l*), (*g*), (*aq*). Assume that the water in the nitric acid solution does not take part in the reaction.

1. Which two of the *products* must be in the solution in the test tube? Describe the appearance of each one. What happened to the copper atoms that were originally part of the wire? Where did they go?

2. What is the name and formula of the yellowish gas you saw? What elements are in the yellowish gas? Where did they originate? (From which reactant did they come?)

3. Where did the atoms that formed the water come from (which reactant)?

4. The process that occurred in your test tube was a chemical reaction. Consider how your teacher cleaned the piece of copper wire before the experiment. What happened to the atoms on the surface of the wire? Was this a chemical reaction? Explain.

5. The equation that precedes Question 1 has been copied below. As you see, the same elements are represented on both sides; that is, there are the same kinds of atoms in the reactants as there are in the products: copper, hydrogen, nitrogen, and oxygen. Notice, however, that the *number* of atoms of each element is not the same. Demonstrate the imbalance by completing this atom inventory. Three of the numbers have been supplied to help you get started.

$$Cu(s) + HNO_3(aq) \rightarrow Cu(NO_3)_2(aq) + NO_2(g) + H_2O(l)$$

Reactants: Copper atoms- **1**   **Products:** Copper atoms-

    hydrogen atoms-   hydrogen atoms-

    nitrogen atoms-   nitrogen atoms-

    oxygen atoms- **3**   oxygen atoms- **9**

6. The equation for the reaction between copper metal and nitric acid has been rewritten here, but with one small change: the equation is *balanced*. In other words, not only are the same elements on both sides of the equation, the number of atoms of each element is the same on both sides. Complete this second atom survey, similar to the one in Question 5, to show that the equation is now balanced.

$$Cu(s) + 4\,HNO_3(aq) \rightarrow Cu(NO_3)_2(aq) + 2\,NO_2(g) + 2\,H_2O(l)$$

Reactants: Copper atoms-   **Products:** Copper atoms-

    hydrogen atoms-   hydrogen atoms-

    nitrogen atoms-   nitrogen atoms-

    oxygen atoms-   oxygen atoms-

7. One way to define a chemical reaction is to say that it is a process in which the reactant molecules are taken apart, then their atoms are regrouped in a different fashion. Does the reaction between copper and nitric acid fit this definition? Explain.

8. Refer back to your observations. List the ones that demonstrate that a chemical change takes place during this experiment. There are at least four that you should identify. Be specific: don't just say that there was a color change, describe the nature of the change and what it was that underwent the change.

9. During the course of this experiment, the piece of copper wire eventually disappears into the solution. What evidence do you have that this disappearance is more than simply dissolving, such as a sugar cube would do in a glass of water?

10. Look back at your prediction in Prelaboratory Question 2. As it turned out, a gas was formed but there were no bubbles in the beaker of water. How do you explain this?

Experiment 28

Conservation of Mass

Problem
Is mass conserved in a chemical reaction?

Introduction
In everyday life it sometimes seems that the law of conservation of mass is not true. Think about burning a log. At the beginning of the chemical reaction there is a whole log. Yet at the end of the fire the log is replaced by just those small ashes. It certainly doesn't seem that mass is conserved. Can you think of other examples where mass seems to appear or disappear in a chemical reaction? List several before coming to lab.

Scientists investigate chemical reactions by carefully containing all of the materials in the beginning substances (the reactants) and also in the ending substances (the products). In this experiment we will investigate if mass really is conserved when a chemical reaction takes place.

Prelaboratory Assignment
✓ Read the **Introduction** and **Procedure** before you begin.
✓ Answer the Prelaboratory Questions.

1. List three examples of situations where mass seems to appear or disappear during a process.

2. What are the reactants in this experiment?

Materials
Erlenmeyer flask (125-mL)
Small test tube (13 x 100)
Rubber stopper to fit the flask
Safety goggles
Lab apron

Graduated cylinder (25 or 50-mL)
Lead nitrate solution
Potassium iodide solution
Balance

Safety
1. Wear laboratory safety goggles and a lab apron at all times in the laboratory.
2. Lead compounds are toxic. Avoid contact with skin and clothes. Be sure to dispose of the reaction products into an appropriate waste container.

Procedure
1. Find the mass of the Erlenmeyer flask, the test tube and the rubber stopper collectively. Record this as the mass of the empty system.

2. Measure 20 mL of the potassium iodide solution using the graduated cylinder. Pour the solution into the Erlenmeyer flask.

3. Clean the graduated cylinder thoroughly and rinse it well with distilled water.

4. Measure 10 mL of lead nitrate and pour this solution into the small test tube.

5. Carefully place the test tube containing the lead nitrate solution into the Erlenmeyer flask, being careful not to mix the solutions. Stopper the flask. Your system should look like **Figure 1**.

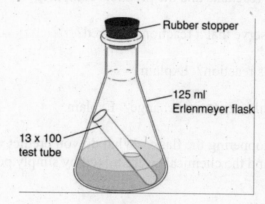

Figure 1
Reaction set-up

6. Record the appearance of each of the solutions.

7. Find the mass of the system containing the two solutions. Record this mass in your table as the mass before mixing.

8. With your hand keeping the stopper in place, carefully tip the system so that the solution in the test tube mixes with the solution in the flask. Record your observations.

9. Find the mass of the system once again. Record this as the mass after mixing.

10. Remove the test tube from the system, pour the contents of the flask into the filter paper provided by your teacher. Try to get as much of the solid as possible into the filter.

Cleaning Up

1. Lead compounds are toxic. Place the filter paper containing your solid in the area designated by your teacher.
2. Wash all glassware and return it to its proper place.
3. Wash your hands thoroughly before leaving the laboratory.

Analysis and Conclusions

Complete the **Analysis and Conclusions** section for this experiment either on your Report Sheet or in your lab report as directed by your teacher.

1. Find the mass of the reactants: (lead nitrate solution + potassium iodide solution).

2. Find the mass of the products after mixing.

3. How did the masses of the reactants and the products compare?

4. What evidence did you observe that a reaction occurred?

5. Was mass conserved in this reaction? Explain.

6. Was this reaction a physical or chemical change? Explain.

7. What was the reason for stoppering the flask? What do you predict would happen if the experiment was repeated and the chemicals were mixed by simply pouring the test tube contents into the flask?

Something Extra

For what kind of chemical reaction would it become essential to stopper the flask in order to show conservation of mass?

Experiment 29

Reactions in Solution I: Precipitation

Problem

What sort of driving forces cause reactions to occur in aqueous solutions?

Introduction

This is a quick and simple experiment that will introduce you to the concept of *driving forces*. By driving force, we mean the reason reactions occur. One of the chemist's most intriguing problems is to understand why things happen as they do. In addition, this experiment will give you some added practice in recognizing and writing equations for precipitation (double replacement) reactions.

You will be given pipets containing ten solutions, all are labeled "0.2 *M*"; this refers to the concentration of the solution. You are to test the solutions in pairs, looking for evidence of reaction in the form of a color change or formation of a precipitate. Remember that on the micro level, since precipitation often amounts to little more than a faint cloudiness it is often best detected by looking down through the well of your test plate at a line of type, such as this one. Some of the pairs will react; many will not. For those that do, write a balanced equation. You will also be expected to identify, by name and formula, the precipitate in each case.

Prelaboratory Assignment

✓ Read the **Introduction** and **Procedure** before you begin.
✓ Answer the Prelaboratory Questions.

1. Write the formulas for the following ionic compounds.
   a. zinc sulfide
   b. chromium(III) hydroxide
   c. lead(II) phosphate
2. Write chemical equations for the following electrolytes dissolving in water.
   a. sodium chloride
   b. copper(II) chloride
   c. iron(III) sulfate
3. Write a balanced molecular equation for the reaction that occurs when a solution of cobalt(II)nitrate reacts with aqueous potassium phosphate, producing a precipitate of cobalt(II) phosphate and aqueous potassium nitrate.
4. All of the reagent solutions initially have a concentration of 0.2 *M*. This means that there is 0.2 mole of dissolved substance per liter of solution. What will be the concentrations after mixing, assuming you use equal volumes of each solution?
5. How will you recognize the formation of a small amount of insoluble solid?

Materials

Apparatus
96-well test plate for performing tests
24-well test plate for holding reagents
Wash bottle and cotton swabs for cleaning 96-well plate
Safety goggles
Lab apron

Reagents   Group A and Group B reagents in microtip pipets

Group A	Group B
Sodium iodide, NaI	Cobalt(II) chloride, $CoCl_2$
Sodium carbonate, Na_2CO_3	Copper(II) chloride, $CuCl_2$
Sodium phosphate, Na_3PO_4	Aluminum chloride, $AlCl_3$
Sodium sulfate, Na_2SO_4	Barium chloride, $BaCl_2$
Sodium hydroxide, NaOH	Nickel(II) chloride, $NiCl_2$

Safety

1. A number of these compounds are toxic. You are using very small quantities so avoid contact with them and wash your hands thoroughly before leaving the laboratory.
2. Safety goggles and a lab apron must be worn at all times in the laboratory.

Procedure

1. Use 4-5 drops of each reagent for each test. Note and record any sign of reaction. Test all possible pair combinations of one group A reagent, with one from group B. Record your observations in a Data Table similar to the one shown below:

Group B→	$CoCl_2$	$CuCl_2$	$AlCl_3$	$BaCl_2$	$NiCl_2$

Group A ↓	
NaI	
Na_2CO_3	Recall that not all pairs will react, and that sometimes the evidence of reaction,
Na_3PO_4	especially formation of a slight precipitate, takes a few moments to appear.
Na_2SO_4	
NaOH	

2. If there are any combinations about which you are in doubt, repeat those tests. It may help to check with other lab groups.

Cleaning Up

1. Return the pipets containing unused portions of the reagent solutions to the proper location.
2. Some of the solutions contain transition metal ions, and should not be washed down the drain. Instead, dump the contents of the well plate onto paper towels in the tray labeled, "Transition Metal Waste."
3. You may need to wash your well plate and clean out some of the wells with a cotton swab. The small amount of metal ions that this will introduce to the drain system is not a concern.
4. Wash your hands with soap and water before leaving the laboratory.

Analysis and Conclusions

Complete the **Analysis and Conclusions** section for this experiment either on your Report Sheet or in your lab report as directed by your teacher.

For those combinations in which the formation of a precipitate was observed, do each of the following:
- ✓ Write the name and formula for the precipitate (remember the principle of electrical neutrality: the total numbers of positive and negative charges must be equal).
- ✓ Write a balanced molecular equation for the precipitation reaction that took place.
- ✓ Write a balanced ionic equation, showing all dissolved species as ions, and all precipitates as molecules.

Something Extra

1. In your ionic equations, you will see that some of the ions in the reactants appear in exactly the same form on the products side of the equation. Ions that are present in a reaction, but which do not take part in the precipitation, are called *spectator ions*. It is sometimes useful to write an abbreviated ionic equation, which can be derived from a complete ionic equation, such as you have written for this experiment. For the reaction between barium chloride and sodium sulfate, the complete ionic equation is:

$$Ba^{2+}(aq) + 2\,Cl^-(aq) + 2\,Na^+(aq) + SO_4^{2-}(aq) \rightarrow BaSO_4(s) + 2\,Na^+(aq) + 2\,Cl^-(aq)$$

If we remove the spectators, $Na^+(aq)$ and $Cl^-(aq)$, we have the net ionic equation:

$$Ba^{2+}(aq) + SO_4^{2-}(aq) \rightarrow BaSO_4(s)$$

For each of the other precipitates that you found in your experiment, write the net ionic equation for the reaction that took place.

2. If there were any combinations that appeared to have a reaction, but in which no precipitate was observed, try to identify which ions were responsible for the reaction but do not attempt to write formulas or equations for the reactions. (Hints: What ions were present? What possible combinations could account for the reaction? Are there other combinations in the experiment where any of those same ion pairs were present, but a reaction did not occur?)

Experiment 30

Activity Series for Metals

Problem

How can we classify metals according to their ability to react?

Introduction

On the periodic table metals and nonmetals are grouped on the basis of their physical and chemical properties. You already know some of the basic differences between the two classes of elements, although not all metals behave alike in all respects. For example, both magnesium and zinc react with hydrochloric acid, but copper does not. In other words, the metals differ in their reactivity; we say that some metals are more *active* than others. In this experiment it will be your task to determine the relative reactivities (or, *activities*) of several metals. The following paragraphs introduce new terms and concepts; read them carefully.

When metals form compounds, they do so by becoming positive ions (*cations*); this involves loss of electrons; a loss of electrons is called *oxidation*. By regaining these electrons, the cation returns to its neutral, elemental form; a gain of electrons is called *reduction*. Because of the differences in the make-up of their atoms, some metals lose electrons easily, while others do so only with difficulty. *Active metals* (ones that form compounds easily) hold their outer electrons very weakly, so it is easy for some other ion or element to take one or more of these electrons from the metal atom's outer energy level. *Inactive metals* are ones that do not react readily to form compounds; they hold onto their electrons more tightly, so are not easily *oxidized*.

When an active metal is brought in contact with the cation form of a less-active metal, one or more electrons transfer from the active metal to the less-active ion. Thus, for active metal X and less-active metal Y, we could expect reactions like the following.

$$X + Y^+ \rightarrow X^+ + Y \quad \text{(one electron transferred)}$$

$$X + Y^{2+} \rightarrow X^{2+} + Y \quad \text{(two electrons transferred)}$$

If, on the other hand, the more active metal is present in ion form, while the less-active one is in the elemental state, nothing will happen.

$$Y + X^+ \rightarrow \text{ NO REACTION}$$

In other words, ions of less active metals (like Y^+ or Y^{2+}, above) can oxidize (take electrons from) the neutral atoms of more active metals (such as X), but not the other way around; ions of the more-active element will not oxidize atoms of the less active element. The ion form of an element that loses electrons easily cannot take electrons away from a neutral atom of an element that holds electrons tightly.

Prelaboratory Assignment

✓ Read the **Introduction** and **Procedure** before you begin.
✓ Answer the Prelaboratory Questions.

1. Identify the following changes as either oxidation or reduction.
   a. $Cu^{2+} \rightarrow Cu$ **b.** $Mg \rightarrow Mg^{2+}$ **c.** $Fe^{2+} \rightarrow Fe^{3+}$ **d.** $2H^+ \rightarrow H_2(g)$

2. Consider the following reaction: $Al(s) + Cr^{3+}(aq) \rightarrow Cr(s) + Al^{3+}(aq)$.
   a. What substance is oxidized?
   b. What substance is reduced?

3. Aluminum metal will react with a solution of copper(II) ions, but copper metal will not react with a solution of aluminum ions. Which metal is more active, copper or aluminum?

Materials

Apparatus
24-well test plate
Forceps or tweezers
Wash bottle
Safety goggles
Lab apron

Reagents
$CuCl_2(aq)$
$FeCl_3(aq)$
$Mg(NO_3)_2(aq)$
$SnCl_2(aq)$
$Zn(NO_3)_2(aq)$
Small pieces of Cu, Fe, Mg, Sn, Zn

Safety

1. Wear safety goggles and a lab apron at all times in the laboratory.
2. While none of the materials used in this experiment poses a significant hazard, good laboratory practice should be followed.

Procedure

Six metals will be tested for their relative activities. The metals to be compared are copper, iron, magnesium, silver, tin, and zinc. Silver is quite expensive, so it will be tested only by your teacher as part of a class demonstration following the experiment; you will test the other five yourself. You will do the tests by placing samples of the metals in solutions containing the cation form of each of the other metals. All the metal pieces have been freshly cleaned of any oxide surface coating.

1. Put a single piece of one of the metals in each of four different wells of your 24-well test plate, then add each other metal solution on separate pieces of the metal being tested, noting cases in which a reaction occurs. Use just enough of each solution to cover the metal piece--8 to 10 drops should be plenty. Quite often, the only sign of reaction will be a darkening of the metal surface, so careful observation is a must. Remember that some reactions are slow to occur, so you should wait at least four or five minutes before you decide that nothing happened.

2. Proceed in this way until all five metals have been tested with solutions of each of the others. Record all of your observations in a Data Table.

Cleaning Up

1. You will find a large pan lined with paper towels. Shake the contents of your test plate onto the towels. Use a pair of forceps or tweezers to pick out pieces of metal that don't shake out, then use a wash bottle to rinse solutions from the test plate into the sink. **Thoroughly clean out the wells** of your test plate; once a piece of oxidized metal or an ion solution has had a chance to dry out, cleaning becomes almost impossible.
2. Wash the test plate, rinse with tap water then with distilled water. Leave the plate face-down on a folded paper towel to dry.
3. Wash your hands before leaving the laboratory.

Analysis and Conclusions

Complete the **Analysis and Conclusions** section for this experiment either on your Report Sheet or in your lab report as directed by your teacher.

1. The metal that reacted with the greatest number of cation solutions is the most active metal you were testing; which one is it?

2. The metal that reacted with the fewest cation solutions is the least active; which is it?

3. Rank the remaining three metals, so that you end up with a list of the five metals in order of decreasing activity (most active first). Use the form, (most) > (next) > , etc.

4. Your teacher demonstrated the reaction that occurs when copper metal is placed in a solution of silver ions. Use your observations to place silver in its proper spot on your activity list.

5. Recalling that the more active metals are those that are more easily oxidized, where on the periodic table would you expect to find the most active metals? Are the more active metals found on the left, the right, or in the middle of the table? What about vertically: are the metals near the top of the table more reactive or less reactive than those toward the bottom?

6. For those cases in which a reaction was observed, write the equation for that reaction. Remember that the number of electrons lost by one kind of atom must equal the number gained by the other, so the total charge on the left side of the arrow must be the same as the total charge on the right side.

Something Extra

Consult a table showing the accepted order of metal activities. Your teacher will either suggest a suitable reference or provide you with the list. Discuss your findings.

Experiment 31

Unknown Solutions

Problem
How can you use chemical reactions and physical properties to identify the unknown solutions?

Introduction
You have come a long way in your knowledge and understanding of chemistry. Now you are ready to use this knowledge and understanding to identify ten unknown solutions.

Prelaboratory Assignment
✓ Read the entire experiment before you begin. Review previous labs, class notes, and the text to find distinguishing characteristics for each of the solutions.
✓ Answer the Prelaboratory Questions.

1. Write the names and formulas for all of the substances in the solutions you will be using in this experiment.

2. List methods to test the solutions. (**Hint:** Don't forget that you can react the solutions with each other).

Materials
Apparatus
Safety goggles
Lab apron
Well plate (24 wells)
Pipets
Bromthymol blue (an acid-base indicator)
Conductivity tester
Reagents
10 unknown solutions

table sugar	silver nitrate
potassium nitrate	sodium hydroxide
sulfuric acid	barium nitrate
copper (II) nitrate	ammonium chloride
sodium chloride	acetic acid

Safety
1. Safety goggles and a lab apron must be worn at all times in the laboratory.
2. Unknown chemicals can be hazardous. If you come in contact with any solution, wash the contacted area thoroughly.

Procedure

1. Discuss the procedures to test the solutions you have developed in the **Prelaboratory Assignment** with your partner and your teacher.

2. Carry out the procedures. Record your procedures and observations.

Cleaning Up

1. Empty the well plate onto a paper towel in the container specified for chemical waste.
2. Rinse the well plate thoroughly with tap water, then with distilled water. Place it upside down on a paper towel to drain.
3. Wash your hands thoroughly before leaving the laboratory.

Analysis and Conclusions

Complete the **Analysis and Conclusions** section for this experiment either on your Report Sheet or in your lab report as directed by your teacher.

1. List each observation and state the possibilities from each observation.

2. Make a table that includes the name and formula of each unknown, along with an explanation of your reasoning for each choice.

Something Extra

Suppose you were given an unknown solution containing a mixture of two or more of the solutions used in this experiment. Describe what you could do to identify the components of this unknown solution.

Experiment 32

Qualitative Analysis: Anions

Problem

How can we identify which anions are in an aqueous solution?

Introduction

Questions like the one above can be answered using the techniques of **qualitative analysis**. The word *qualitative* refers to the fact that you are simply trying to ascertain whether a particular ionic species is present. You are not trying to find out how much is present. The presence of an ion is usually signaled by formation of a precipitate or by a color change. In this experiment, the tests are similar to ones you have seen before, but in some cases the results are more subtle and there are more possibilities to work with, so identification of unknowns may be more of a challenge. The tests all consist of observing how the anions of interest respond to particular metal cations.

Since it isn't possible to have a solid or solution that contains only anions, the solutions you will be using will contain both a cation and an anion. But the cation in each case (sodium) is a non-participant in any reaction that you observe. Therefore, any reaction that occurs can be assumed to be between the anion you are trying to identify and the test cation. Similarly, each of the cation solutions: iron(II), iron(III), and silver has been paired with an anion that will not become involved in any reaction that takes place.

Synopsis: Testing will consist of adding solutions containing Fe^{2+} (with and without acid), Fe^{3+}, and Ag^+, to solutions containing each of ten anions. For each test you must decide whether you see cloudiness, indicating the formation of an insoluble precipitate, a color change that is not simply the result of mixing or dilution, or no reaction at all.

A word about color changes: If you add a colorless solution to one that is yellow, having the yellow color grow more pale is expected. If, on the other hand, the yellow turns to green, that shows a new species has been produced, so a reaction has occurred.

Here is a list of common anions for which you will be testing:

Halide ions:	bromide, (Br^-); chloride, (Cl^-); iodide, (I^-)
Polyatomic anions:	carbonate, (CO_3^{2-}); hydroxide, (OH^-); oxalate, ($C_2O_4^{2-}$); nitrate, (NO_3^-); phosphate, (PO_4^{3-}); sulfate, (SO_4^{2-}); and thiosulfate, ($S_2O_3^{2-}$).

Prelaboratory Assignment

✓ Read the **Introduction** and **Procedure** before you begin.
✓ Answer the Prelaboratory Questions.

1. Sulfuric acid is used in this experiment. Describe the process for cleaning up a small amount of sulfuric acid that has spilled on the desktop.

2. For each of the following, indicate whether there is evidence of a chemical change.
   a. Mixing two clear, colorless solutions causes formation of a bright yellow solid.
   b. Mixing two clear, colorless solutions produces some bubbles which eventually disappear, leaving a clear, colorless solution.
   c. Mixing two clear, colorless solutions initially gives a clear, colorless solution. As it stands, the solution turns hazy.
   d. Addition of a white solid to a clear, colorless solution produces some bubbles that disappear after about 5-10 seconds, leaving a clear, colorless solution.

Materials

Apparatus
Safety goggles
Lab apron
24-well test plate (one or two)
Distilled water wash bottle
Cotton swabs for cleaning

Reagents
Halide ion test solutions:
$NaBr(aq)$
$NaCl(aq)$
$NaI(aq)$

Polyatomic anion test solutions:
$Na_3PO_4(aq)$
$Na_2C_2O_4(aq)$
$Na_2CO_3(aq)$
$Na_2S_2O_3(aq)$
$NaNO_3(aq)$
$Na_2SO_4(aq)$
$NaOH(aq)$

Test reagents:
Iron(II) solution: $FeSO_4(aq)$
Iron(III) solution: $Fe(NO_3)_3(aq)$
Silver ion solution: $AgNO_3(aq)$
Sulfuric acid, H_2SO_4, 8.0 M

Unknowns: (3 samples, selected from the list of anion solutions tested)

Safety

1. 8.0 M sulfuric acid is extremely hazardous. Immediately flush any spills with large amounts of water.
2. Silver compounds are very toxic. Wash your hands thoroughly with soap and water before leaving the laboratory. At no time is food, drink, or chewing gum allowed in the laboratory. Toxic contamination poses a special hazard in this case.
3. Always wear safety goggles and a lab apron in the laboratory.

Procedure

If you are not using a Report Sheet for this experiment design a Data Table with the following headings:

Anion	Result with Fe^{2+}	Effect of H_2SO_4	Result with Fe^{3+}	Effect of H_2SO_4	Result with Ag^+

Detailed instructions for testing the sodium bromide solution appear below. Follow the same procedure for each of the remaining anions. Once you have completed the process with bromide follow the instructions in **Cleaning Up**, then resume testing until the sequence has been completed with all 10 anions.

Because silver salts are quite toxic, the tests with silver ion will have to be done separately. Do not rinse the results of the silver tests down the drain. Dispose of them as directed by the teacher.

Part 1   Tests with Iron(II) and Iron(III) Ions

1. Into two separate wells of a 24-well test plate, place 10 drops of $NaBr(aq)$, (sodium bromide solution). To one of the wells, add 5 drops of iron(II) sulfate, ($FeSO_4(aq)$). Observe and record any changes that occur.

2. If no changes are observed, add 3 drops of 8 M sulfuric acid, H_2SO_4, and observe again. **Note:** There are two types of changes that you must look for when adding strong acid: (1) evolution of a gas; (2) a color change that is not the result of mixing or dilution.

3. To the second well containing NaBr solution, add 5 drops of iron(III) nitrate, ($Fe(NO_3)_3(aq)$). As before, observe and record any changes that occur. If no changes are observed, add 3 drops of 8 M sulfuric acid, and observe and record any evidence of a reaction.

4. Repeat steps 1-3 for chloride, iodide, and all the remaining anions. As often as you need to, follow the instructions for Part 1 under **Cleaning Up**.

Part 2   Tests with Silver Ion

1. In a clean well plate, duplicate the tests done with iron(II) and iron(III), using instead, silver nitrate, ($AgNO_3(aq)$). You need not add acid to these tests. Note and record the nature of any reactions that take place. Be sure to test all 10 of the anion solutions.

Part 3   Identification of Unknowns

1. You will be given three unknown solutions, one at a time. Complete your tests for one before you start the next.
   - The first unknown contains one of the 3 halide ion solutions listed in the Materials section.
   - The second will contain either iron(II), iron(III) or silver ion.
   - The third will contain one of the 7 polyatomic anion solutions.

   For each unknown, briefly describe what you did, what you observed and what information you got from that test. Remember that "no reaction" can be just as useful as appearance of a precipitate, if it allows you to eliminate some possibilities.

Cleaning Up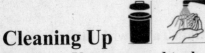

Part 1 You may need to do this more than once, depending on whether you find it necessary to repeat any tests.
1. Clean the well plate thoroughly. Rinse it well, then use a cotton swab to remove all traces of precipitates from the wells. Rinse with distilled water and shake it dry.
2. If the same well plate is to be used for the silver tests, continue on to Step 5. If the silver tests are to be done with a different well plate, return all materials to their proper location and return to the classroom or proceed to **Part 2, Identification of Unknowns**.

Part 2 Disposing of Silver Ion:
1. To clean the well plate containing silver ion solutions, first, shake the contents of the wells onto the paper towels in the tray indicated by the teacher. Then use a wash bottle and cotton swabs to remove any materials that did not come out with shaking. Finally, rinse the well plate once more with distilled water (these rinsings can go down the drain) and return all materials to their proper location.

Part 3 Disposal for unknowns
1. If silver ion was used to help identify the unknowns, follow the method for Part 2.
2. Clean the well plate thoroughly. Rinse it well, then use a cotton swab to remove all traces of precipitates from the wells. Rinse with distilled water and shake it dry.
3. Wash your hands with soap and water before leaving the laboratory.

Analysis and Conclusions

Complete the **Analysis and Conclusions** section for this experiment either on your Report Sheet or in your lab report as directed by your teacher.

1. Suppose you were given three unlabeled solutions. You know that one contains iron(II), another contains iron(III) and the third has silver ion, but you don't know which is which. Devise a scheme by which you could identify each of the three cation solutions.

2. Explain how you can distinguish among solutions containing bromide, chloride and iodide ions.

3. Explain how you can determine whether oxalate ion is present in a solution.

Experiment 33

The Halide Family

Problem
What tests will enable us to identify and distinguish between ions in the halide family?

Introduction
You are familiar with the fact that elements in the same vertical column on the periodic table share similar behavior. The key word here is "similar." Members of the same family can be identified by differences in physical properties: density, melting point, molar mass, and so on. They can also be distinguished by differences, sometimes slight, in their chemical behavior. This is especially useful when we are dealing with the elements in the form of dissolved ions.

The term *halide ion* refers to any of the ions, F^-, Cl^-, Br^-, or I^-. In this experiment you will focus specifically on this family of ions. You will be given solutions containing known halide ions. Then you will use several other solutions (*test reagents*) to see how each reacts with a particular halide ion. By carefully observing how each halide ion reacts, you should be able to identify each halide as an unknown and subsequently identify two halide ions mixed together in solution.

Prelaboratory Assignment
✓ Read the **Introduction** and **Procedure** before you begin.
✓ Answer the Prelaboratory Questions.

1. What is a *precipitate*?
2. At one point in the procedure, you will try two different tests on the same precipitate. Describe the process for dividing the precipitate into two separate samples.
3. Write equations for each of the following processes.
   a. Sodium thiosulfate dissolves in water.
   b. A precipitate forms when solutions of silver nitrate and sodium iodide are mixed.

Materials
Apparatus
96-well test plate
Toothpicks(several)
Microtip pipet for mixing
Distilled water wash bottle
Cotton swabs
Safety goggles
Lab Apron
Gloves (recommended)

Test reagents:
Calcium nitrate, $Ca(NO_3)_2$ (*aq*)
Silver nitrate, $AgNO_3$ (*aq*)
ammonia solution, NH_3 (*aq*)
sodium thiosulfate, $Na_2S_2O_3$ (*aq*)
Starch solution
Bleach

Reagents
Halide ion solutions:

sodium fluoride, NaF (*aq*)

sodium chloride, NaCl(*aq*)

sodium bromide, NaBr (*aq*)

sodium iodide, NaI (*aq*)

Optional Reagents:

Hexane, $C_6H_{14}(l)$ (dropper bottle)

Chlorine water, Cl_2 (*aq*)

Safety

1. Wear safety goggles and a lab apron. Gloves are recommended for this experiment.
2. Silver nitrate is toxic and will stain skin and clothing. Wash your hands thoroughly with soap and water before leaving the laboratory. At no time is food, drink, or chewing gum allowed in the laboratory. Toxic contamination poses a special hazard in this case.
3. Ammonia is caustic and has a strong, pungent odor. It will irritate skin and is particularly dangerous to the eyes.

Procedure

If you are not using a Report Sheet for this experiment design a Data Table with the following headings:

Anion	Reaction with Ca^{2+}	Reaction with Ag^+	Response to starch	Response to bleach

Part 1

1. Obtain labeled thin-stem pipets containing solutions of NaF, NaCl, NaBr, and NaI.

2. Add 4 drops each of NaF, NaCl, NaBr, and NaI to four separate wells of your well plate.

3. Add 4 drops of calcium nitrate, $Ca(NO_3)_2$, to each well and observe.

Part 2

1. Add 4 drops each of NaF, NaCl, NaBr, and NaI to four separate wells of your well plate.

2. Add 4 drops of silver nitrate, $AgNO_3$, to each well and observe (note both the presence of any precipitates and their colors).

3. Split the precipitates, keeping track of which halide ion is contained in which well. First, stir the precipitate by placing the tip of a clean microtip pipet on the bottom of the well then use a gentle squeeze of air to stir the mixture. Take everything up into the pipet and then distribute it drop by drop between the original well and another one next to it.

4. Add 4 drops of NH_3(aq) to one half of each precipitate, stir with a toothpick, and observe. Does ammonia cause the precipitate to dissolve? Discolor?

5. Add 4 drops of sodium thiosulfate, $Na_2S_2O_3$, to the second half of each precipitate, stir with a toothpick, and observe. Is there any effect on the precipitate?

Part 3

1. Add 4 drops each of NaF, NaCl, NaBr, and NaI to four separate wells of your well plate.

2. Add 2 drops of starch solution to each well. Stir and observe.

3. Add 2 drops of commercial bleach (NaOCl) to each well, stir, and observe again.

Part 4   (Optional; do this part only if instructed to by your teacher.)

1. If further confirmation is desired, add 4 drops of NaF, NaCl, NaBr, and NaI separately to each of four small vials or test tubes.
   Note: Plastic well-plates are not suitable for this part of the experiment.

2. Add 4 drops of hexane to each solution and observe.

3. Add 2-5 drops of chlorine water to each tube, stir or shake gently, and observe.

Part 5   Unknowns

For each of your unknowns, indicate the tests that allowed you to identify the ion or ions present. Include what reagents you used in the tests and what you observed.

1. Obtain an unknown halide ion solution and test it with each reagent or combination of reagents as indicated in Parts 1 through 3 until the identity of the halide ion present has been determined.

2. Obtain an unknown solution which contains two of the four halide ions, mixed together. Test with each reagent or combination of reagents until the identities of both halide ions are known.

Cleaning Up

Parts 1-3 and Part 5:

1. Shake the contents of the test plate onto paper towels in the tray provided for that purpose. Use soap and water and cotton swabs to remove any precipitates from the wells. Rinse the tray with distilled water. If you do not need to use it again, leave it to dry on paper towels. If you need to use it further, shake it as dry as you can. A dry cotton swab may help dry the wells.

Part 4:   The tubes containing hexane and molecular (elemental) halogens, both require special disposal. If you are completing the Something Extra be sure to record observations during Cleaning Up.

1. To convert any molecular halogens to the corresponding halide ion, add a few drops of sodium thiosulfate solution to each tube and mix.
2. Take the tubes to the fume hood and pour the contents into the container labeled **organic waste.** Your teacher will handle disposal from there.
3. Wash your hands thoroughly before leaving the laboratory.

Analysis and Conclusions

Complete the **Analysis and Conclusions** section for this experiment either on your Report Sheet or in your lab report as directed by your teacher.

1. Explain what information you found from each of the first three (or four) parts of this experiment. Make separate clear, concise and grammatically correct statements for each part. Someone else should be able to use your results to analyze the unknowns successfully.

2. In what way does fluoride ion behave differently from the other three ions in the family?

3. In what way(s) do chloride, bromide, and iodide ions behave similarly? In what way(s) do they display different chemical behavior?

4. How can you distinguish between:
    a. chloride ion and bromide ion?
    b. chloride ion and iodide ion?
    c. bromide ion and iodide ion?

Something Extra

1. If you did the optional tests from Part 4, describe and account for the color changes you observed during **Cleaning Up**.

2. A common test for the presence of starch is formation of a dark, blue-black color when iodine, I_2, is added. How do you account for the color you observed in Part 3? What must have happened?

Experiment 34

Copper Wire
in a Solution of Silver Nitrate

Problem

What mass and mole relationships are found in the reaction between copper and silver nitrate? What chemical changes occur when a metal sample is reacted with a metal compound?

Introduction

In this experiment you will observe the chemical changes that occur when copper reacts with silver nitrate. You will also make careful measurements of the masses of copper and silver nitrate that react. While it is convenient for us to measure the *mass* of substances that react in a particular process, this experiment will help show that reactions involve rearrangements of atoms. Thus the equations that describe chemical reactions involve the number of atoms that participate. You have a chance in this experiment to find the numerical relationship between copper and silver nitrate when they react.

Prelaboratory Assignment

✓ Read the **Introduction** and **Procedure** before you begin.
✓ Answer the Prelaboratory Questions.

1. Write the unbalanced equation for the reaction between copper and silver nitrate.
2. How many atoms are there in a 10.4 g sample of copper?
3. How many $AgNO_3$ units are there in a 2.2 g sample of silver nitrate?
4. In another experiment 1.65 grams of lead nitrate, $Pb(NO_3)_2$, react with 0.327 g of zinc, Zn. What is the mole relationship between the lead nitrate and the zinc?

Materials

Apparatus
Safety goggles
Lab apron
Gloves (optional)
100 or 150-mL beaker
Glass stirring rod
Wash bottle
250-mL beaker
Parafilm™
Centigram or milligram balance

Reagents
Copper wire
Steel wool
Vial containing silver nitrate crystals
Distilled water
Acetone

Safety

1. Wear safety goggles and lab aprons at all times in the laboratory. Gloves may be used to protect your hands when using the silver nitrate.
2. Be careful not to touch the silver nitrate crystals or the silver nitrate solution with your hands. The silver nitrate will stain your skin black. Wash your hands thoroughly with soap and water before leaving the laboratory. At no time is food, drink, or chewing gum allowed in the laboratory. Toxic contamination poses a special hazard in this case.

Procedure

Day 1   Starting the Reaction

1. Obtain a length of copper wire. Use a piece of steel wool to polish and clean it.

2. Find the mass of the copper wire. Find the mass of a clean, dry, labeled 100- or 150-mL beaker. Find the mass of a vial of silver nitrate crystals.

3. Transfer the crystals of silver nitrate to the weighed 100- or 150-mL beaker. Find the mass of the empty vial.

4. Add 75-80 mL of *distilled water* to the silver nitrate crystals. Swirl gently until the crystals dissolve. **Note:** Be careful not to touch the solution of silver nitrate with your hands; the silver nitrate will stain your skin black.

5. Coil the copper wire so that most of the wire is exposed to the solution when it is placed in the beaker. Keep about 3-4 cm of wire above the solution to serve as a handle for removing the copper wire later.

6. Place the copper wire in the silver nitrate solution. Observe the contents of the beaker for several minutes and record any changes that occur.

7. Cover the beaker with Parafilm™ and place it in the assigned spot until the next laboratory period.

Day 2   Observing and Isolating the Products of the Reaction

1. Carefully place the beaker on your lab bench. Observe what has happened in the beaker. Record all observations.

2. Carefully shake the copper wire to remove the crystals. Try to dislodge as many crystals as possible by using your wash bottle of distilled water or the tip of a glass stirring rod.

3. Place a small amount (~10 mL) of acetone in a small beaker. Dip the copper wire into the acetone and set it aside to dry for several minutes. When it is dry, find the mass of the copper wire.

4. Allow the crystals of product to settle in the beaker. Carefully decant the solution into a 250-mL beaker. Be sure to keep all of the crystals in the 150-mL beaker.

5. Wash the residue in the 150-mL beaker with about 10 mL of distilled water. Decant once again into the larger beaker. Wash and decant at least three more times.

6. After the final washing the residue must be dried. Your teacher will instruct you in the proper method for doing this.

Cleaning Up
1. Place the solution from the 250-mL beaker in the waste container provided by your teacher.
2. Clean up your equipment and return it to its proper location.
3. Wash your hands before leaving the laboratory.

Day 3   Measuring the Final Product
1. Find the mass of the cool, dry 150-mL beaker and residue.

Cleaning Up
1. Place the residue in the container provided by your teacher.
2. Clean your beaker and return the glassware to its proper location.
3. Wash your hands before leaving the laboratory.

Analysis/Conclusions

Complete the **Analysis and Conclusions** section for this experiment either on your Report Sheet or in your lab report as directed by your teacher.

1. Calculate the change in the mass of the copper wire.

2. Calculate the mass of silver nitrate used.

3. Calculate the mass of residue (silver) obtained.

4. Determine the number of moles of copper that reacted.

5. Determine the number of moles of silver nitrate used in the reaction.

6. Determine the number of moles of silver produced in the reaction.

7. From your data, determine the ratio: $\dfrac{\text{moles Ag}}{\text{moles Cu}}$

   Express your answer as a decimal to the proper number of significant digits.

8. From your data, determine the ratio: $\dfrac{\text{moles Ag}}{\text{moles AgNO}_3}$

9. Summarize your results in a Summary Table comparing the mass and moles of each reactant and the silver product.

10. Write a balanced chemical equation for the reaction between copper and silver nitrate.

11. How many atoms of copper reacted in this experiment? How many atoms of silver were produced?

12. How do the ratios that you calculated in 7 and 8 above relate to the balanced equation?

13. What caused the color that you observed in the 150-mL beaker as the experiment proceeded?

14. Why was the mass that the copper lost not equal to the mass of the silver that formed?

Something Extra

1. The values obtained by the class for the Ag:Cu ratio and for the $Ag:AgNO_3$ ratio will be listed on the board. Use the posted information to calculate the average ratios for the class.
   a. How do you explain a Ag:Cu ratio which is higher than the average? Lower than the average?

   b. How do you explain a $Ag:AgNO_3$ ratio which is higher than the average? Lower than the average?

Experiment 35

Mass Relationships in Chemical Compounds

Problem
How can we verify the Law of Constant Composition?

Introduction
When John Dalton formulated his atomic theory of matter in the early 1800's, he used the Law of Constant Composition as the basis for proposing that all matter consisted of atoms. In this experiment you will start with an unknown compound. Your task will be to observe and record changes in appearance and mass that accompany a series of reactions. The calculations you will need to do are simple if you collect the proper data. Remember to record all observations and data. Because the activity will extend over parts of three class periods, it is important to label all containers and samples that are stored. Record the date each time you resume work.

Prelaboratory Assignment
✓ Read the **Introduction** and **Procedure** before you begin.
✓ Answer the Prelaboratory Questions.

1. Of the types of reactions discussed in Chapters 7 and 8 of your text, which type would most likely begin with a single reactant and involve an apparent decrease in mass.
2. Why must you avoid using the stirring rod to break up the solid while it is in the filter?
3. Discuss the effects that might result from the following typical student technique errors. Specify whether the mass of the sample will appear to be greater or less than it should be.
   a. The student forgets to find the mass of the empty beaker before beginning, and uses the mass of another beaker in the calculations.
   b. The student finds the mass of the empty beaker, then adds a masking-tape label for ease of identification later.
   c. The student uses a stirring rod to break up the product, then places the rod on the desk top between uses.
   d. To save time, the student finds the mass of the filter paper and product immediately after filtration, rather than waiting for it to dry overnight.

Materials

Apparatus
Milligram or centigram balance
30-mL or 50-mL beaker
Hot plate
Stirring rod
Forceps or tweezers
Funnel
Filter paper

Reagents
Unknown blue compound
Aluminum strip, about 5 cm x 0.5 cm
Distilled water, in wash bottle

125-mL Erlenmeyer flask
Tongs (for handling beaker)
Small spatula
Safety goggles
Lab apron

Safety

1. Wear safety goggles and a lab apron at all times in the laboratory
2. Use caution when handling the hotplate and hot glassware.
3. Unknown chemicals can be caustic. Avoid contact with your skin.

Procedure

Part 1:   Day One

1. Find the mass of a labeled, dry 30- or 50-mL beaker. Add between 0.3 and 0.5 g of the unknown blue compound. Record the masses to the nearest milligram (0.001 g) in your data table.

2. Using tongs to hold the beaker steady, gently heat the beaker and sample on a small hot plate. Stir constantly with a stirring rod to prevent charring of the solid. (charring will appear as a black area in the mass of solid). Continue until the entire sample has changed color. The heating process should take less than one minute. Scrape any grains from the stirring rod back into the beaker. Record your observations.

3. Use the tongs to remove the beaker from the hot plate. Set the beaker on a wire screen and allow it to cool for at least 10 minutes, or until it is at room temperature, then determine and record the total mass. **Note:** Do not put a hot object on the balance. You may damage the balance and your measurement will be incorrect.

4. Beginning with a few drops at a time, slowly add about 15-20 mL of distilled water to the beaker. Describe all changes in the appearance of the sample as the water is added.

5. Obtain a strip of aluminum metal, shape it into a loose coil, and place it into the solution from step 4. Observe all changes that occur for the first few minutes. Cover the beaker with a watch glass or a weighted piece of filter paper and place it in the location indicated by your instructor until the next laboratory period. Your notes should describe both the changes that occur as the reaction is starting and the appearance of the system the next day.

Part 2:   Day Two

6. Use tweezers or forceps to remove any aluminum that remains in the beaker. Use a spatula to scrape off any solid clinging to the aluminum into the beaker. Then rinse and discard the piece of aluminum in the container labeled *aluminum waste*. Because the solution is acidic, rinse your hands well immediately after you finish.

7. Weigh a piece of filter paper, fold it into a cone, and place it in a glass or clear plastic funnel that is suspended in a 125-mL Erlenmeyer flask. Now pour the *liquid only* from the beaker into the filter, retaining as much as you can of the solid product in the beaker. Allow the liquid to drain through the filter, then rinse the filter lightly with distilled water from a wash bottle.

8. Add about 10-15 mL of distilled water to the solid in the beaker. Use a stirring rod to break up the solid in the water. Drain the liquid through the filter as before, then repeat the rinsing (in order for all contaminants to be removed, the rinsing must be thorough).

9. Using a clean spatula, completely transfer your solid product from the beaker to the filter. Use a wash bottle to rinse away any remaining impurities from your product. Drain the rinsings, then repeat. When all rinsings have drained through the filter, carefully remove the filter with your product from the funnel. Place it on a labeled paper towel in a location where it may be safely left to dry.

Part 3:   Day Three
10. Observe and record the appearance of the dry product. Find and record the mass of the dry filter paper and product, then dispose of them as directed by your teacher.

Cleaning Up

1. All solutions may be safely rinsed down the sink with large amounts of water.
2. Wash your hands with soap and water before leaving the laboratory.

Analysis and Conclusions

Complete the **Analysis and Conclusions** section for this experiment either on your Report Sheet or in your lab report as directed by your teacher

1. Calculate the mass of the original sample, the mass lost during heating and the mass of solid remaining in the beaker.

2. Calculate the mass of the final product.

3. Calculate the ratio of the mass of the final product to the mass of the original blue solid.

4. Calculate the ratio of the mass of final product to the mass of the solid remaining after heating.

5. Given what you observed when you added distilled water in step 4, suggest a reasonable explanation for the mass loss during heating.

6. What might have happened if you had left the beaker and contents from step 3 overnight before finding the mass.

7. Cite at least three examples of evidence for chemical change that you observed in step 5. Of the categories of reactions discussed in your text, which type was involved in step 5 of the procedure? How do you know?

8. Does your final product appear to be the same substance as either of the other solids in this experiment (don't include aluminum in your consideration)? Explain.

9. Compare your ratios from **3** and **4** to those of other teams. Are the results consistent or do they appear to vary in random fashion? Does this experiment illustrate the Law of Constant Composition? Explain.

10. In view of your answers to **5-7**, and given the demonstration your teacher carried out during the post-laboratory discussion, did the heating of the original blue solid result in a chemical change or a physical change. Explain.

Something Extra

Now that you have analyzed the mass ratios involved, it is time to see how well your data extend to include mole ratios. Here is an overview of what has happened in this experiment.

The initial heating of the blue solid simply removed water from a sample of hydrated copper(II) chloride. Many compounds absorb water from the air into their crystalline structure. Here is the equation for the dehydration:

$$CuCl_2 \cdot 2H_2O(s) \xrightarrow{\Delta} CuCl_2(s) + 2H_2O(g)$$

The Δ above the arrow indicates that heat is necessary for the process. Once the brown solid was cooled and weighed, you slowly added water to it. At first, the water simply replaced what had been lost during heating, but as you added more and more water, you dissolved the copper(II) chloride, forming a solution. Addition of aluminum resulted in a single replacement reaction:

$$2Al(s) + 3CuCl_2(aq) \longrightarrow 3Cu(s) + 2AlCl_3(aq)$$

Answer the following, showing your calculations.

1. The last product you obtained (in the filter paper) was copper metal (step 10).
   a. What mass of copper did you recover?
   b. How many moles of copper did you recover?
   c. How many moles of aluminum must have reacted?
   d. What mass of aluminum was consumed?

2. The light brown solid that was left after the heating consisted only of copper and chlorine. Assume that the mass of the copper that you recovered in step 10 was the same as the mass of copper present in the compound of copper and chlorine remaining after step 3.
   a. What mass of chlorine was in the compound?
   b. How many moles of chlorine were in the compound?
   c. What was the mole ratio of chlorine to copper in the compound?

3. The ratio you reported in **2c**, represents the number of chlorine atoms per copper atom in the compound, so it must be a whole number ratio (1/1, 2/1, 3/2, 3/1, etc.)
   a. What whole-number ratio best matches your answer to 2c?
   b. Write the formula for the compound.

4. Copper only forms two cations, Cu^+ and Cu^{2+}; chloride is always Cl^-. Based on your experimental results, did you have a sample of CuCl or of $CuCl_2$? Defend your choice.

5. Starting with the mass of the copper chloride in the beaker after step 3, calculate:
   a. The number of moles of copper chloride you had.
   b. The number of moles and the mass of copper that should be present in your sample of copper chloride. This is your theoretical yield.
   c. Divide the mass of copper actually recovered (step 10) by the theoretical yield you just calculated, then multiply that answer by 100%. This is your percent yield.

6. The initial heating drove off water that had been trapped in the copper chloride sample.
   a. What mass of water was driven off during heating?
   b. How many moles of water were driven off?
   c. In the original blue-green compound, what is the mole ratio of:
      i. water to copper
      ii. water to chlorine

Experiment 36

The Calcium Content of Milk

Problem

How can we determine the calcium content of milk?

Introduction

Calcium has long been known to be necessary for a healthy diet. Dairy products are touted as an excellent source of calcium. In this experiment you will determine the amount of calcium in milk by gravimetric titration with EDTA. The calcium concentration in the milk will be determined by finding the mass of the EDTA solution needed to titrate a known mass of milk to a blue endpoint. This process is similar to the method often used to find the hardness of water which is caused primarily by magnesium and calcium ions.

EDTA (ethylenediamine tetraacetic acid) is a large molecule that has a strong attraction for metallic ions like Ca^{2+}, the form of calcium in milk. The calcium ions and the EDTA combine in a 1:1 mole ratio to form a large complex ion called a "chelate." Chelates, such as the hemoglobin which contains the Fe^{2+} ions and is found in your blood cells, are common in nature.

In this titration, you will carefully weigh a pipet full of EDTA solution, then add the EDTA dropwise to a known mass of milk. An *indicator* will be added to the milk to tell you when all the calcium ions have been removed from the milk. The indicator that you will use, hydroxy naphthol blue, changes from red, when calcium is present, to blue, when all of the calcium ions have been removed from the solution. As you add the EDTA, the calcium ions will be used up, causing the indicator to change color. The blue end point appears gradually, so you may want to run a preliminary trial just to observe the color change before you do your quantitative measurements. Because the indicator only works when the system is basic, you will add a drop of 6-molar sodium hydroxide solution to each of your titration samples.

Remember that red and blue make purple; the mixture will be purple when both the calcium-containing form and the calcium-free form of the indicator are present in about equal amounts. This will occur before the blue endpoint.

Using the mass of EDTA in your solution and the mass of the milk, you will be able to determine the number of milligrams of calcium in 240 mL of milk (about one cup).

Prelaboratory Assignment
✓ Read the **Introduction** and **Procedure** before you begin.
✓ Answer the Prelaboratory Questions.
   1. Why would you expect the calcium in milk to be present as calcium ions, rather than as elemental calcium?

2. According to one source, one cup (~240 mL) of whole milk contains 291 mg of calcium. Use this to calculate the mass percent of calcium in whole milk. Assume the density of milk is 1.0 g/mL.

3. Would you expect the density of milk to be affected by its fat content? Explain. Would increased fat content raise or lower the density? Explain.

Materials

Apparatus
Milligram balance
Thin-stem pipets (3)
24-well test plate
Plastic toothpick for stirring
Safety goggles
Lab apron
Gloves (optional)

Reagents
Milk: whole, 2%, skim
EDTA solution, 1.00%
6 *M* NaOH
Hydroxy naphthol blue indicator

Note: Small beakers or flasks (e.g., 10-20 mL) may be substituted for the 24-well plate.

Safety

1. Sodium hydroxide is caustic, even in small quantities. Clean up spills with large amounts of water.
2. Hydroxy naphthol blue and EDTA are considered irritants; avoid contact with your skin. You may want to wear gloves to protect your hands if you have sensitive skin.
3. Safety goggles and a lab apron must be worn at all times when working in the laboratory.

Procedure

1. Label three pipets: EDTA, NaOH and Milk; fill each with the appropriate solution.

2. Weigh the milk pipet and the EDTA pipet and record their masses in a Data Table. The mass of the NaOH pipet is not needed.

3. Add 15 to 20 drops of milk to one well of your 24 well plate. Reweigh the milk pipet and record its mass in your Data Table.

4. Add one drop of 6*M* NaOH to the milk in the well plate.

5. Add a very small amount of the indicator to the mixture in the well plate. Only a few crystals are needed. Stir the mixture. It should have a red to rose color. If you think the color is too light, add a few more crystals, but if you use too much, results will be inconsistent.

6. Add EDTA solution a drop at a time to the mixture in the well plate. Stir with a toothpick. As you add the EDTA, the calcium ions will be tied up, removing them from the milk solution and causing the color to change, first to purple, then to blue. Record the mass of the EDTA pipet after the solution has turned blue.

7. Carry out two additional trials. If the results do not show good agreement, run additional trials, as needed. If you have not achieved consistent results after 5 trials, consult your teacher.

8. **(Optional)** Determine the density of milk in grams per milliliter, then use this experimental value in your calculations rather than assuming that 240 mL of milk has a mass of 240 g. You must report data and show calculations.

Cleaning Up

1. The 6*M* sodium hydroxide, NaOH, is a strong base, so should be handled with care. Aside from that, there are no environmental concerns connected with this experiment.
2. Depending on the fat content of the milk used, it may take some effort to clean the well plate. The higher the fat content of the milk, the more oily the mixture and the more effort you will need to get the wells clean. Wash everything well with soap and water and return it to the location specified by your teacher.
3. Wash your hands thoroughly before leaving the laboratory.

Analysis and Conclusions

Complete the **Analysis and Conclusions** section for this experiment either on your Report Sheet or in your lab report as directed by your teacher.

1. The concentration of the EDTA solution has been adjusted so that exactly 1 gram of EDTA solution will tie up exactly 1.08 mg of calcium. Calculate the number of milligrams of calcium present in each of your three titration samples. Show your work for the first titration. If you carried out more than three titrations, base all your calculations on the three that show the best agreement.

2. Use your results from the first calculation to determine the number of milligrams of calcium in 1 cup (~240 mL) of milk. Report both the individual values for the three samples and an average value. Assume that skim milk has the same density as water. Show your work for trial 1.

3. Calculate the individual deviations from the average for each trial. Calculate the average deviation for your three trials.

4. When an average deviation is known for a series of analyses, the results of the analysis are generally given in the form: (average value) ± (average deviation). Report the concentration of calcium in milk with the average deviation.

5. Calculate the percent deviation for your experimental results. Show your work.

6. The USRDA for calcium is 1200 mg per day. What fraction of the daily requirement would one cup of milk provide, based on your average value?

7. According to the carton label, one cup of milk provides 35% of the USRDA for calcium. What is your percentage error, assuming the carton value is correct?

8. Women over the age of 50 and men over age of 65 need about 50% more calcium than younger adults. Why is this?

9. A student purchased a calcium supplement tablet which contained calcium carbonate. The student placed the tablet in water for several hours. Addition of NaOH and the indicator gave a blue solution. What does this indicate? Does this test show that the supplement is not giving the student any useful calcium? Explain your reasoning (**Hint:** Consider solubility rules).

Something Extra

Suggestions for further investigation: These are optional, but be aware that it will be up to you, not the teacher, to provide the materials for options 1 and 3.

1. Repeat the experiment, but use powdered milk (reconstituted with water).

2. Try the experiment on a larger scale. Try 5.0-mL samples of milk, using small glass vessels (such as 10-mL Erlenmeyer flasks or 20-mL beakers) in place of the 24-well test plate. Calculate the quantities of other reagents you expect to need and show your calculations to your teacher before you begin.

3. Use a calcium supplement tablet. Consult your teacher; it may be necessary to use acetic acid or some other dilute acid to dissolve the calcium tablet (this is another hint for Question 9, above).

Experiment 37

Stoichiometry

Problem

What is the formula for the iron-containing compound formed when copper (II) sulfate reacts with pure iron?

Introduction

We learned in Chapter 4 that iron can form more than one type of ion. The possible reactions are:

copper(II) sulfate(*aq*) + iron(*s*) → copper(*s*) + iron(II) sulfate(*aq*)

copper(II) sulfate(*aq*) + iron(*s*) → copper(*s*) + iron(III) sulfate(*aq*)

In this lab you will measure the amount of product formed when measured amounts of reactants are mixed. From this information, you will determine which equation applies to the reaction you run in this experiment.

Prelaboratory Assignment
✓ Read the entire experiment before you begin.
✓ Answer the Prelaboratory Questions.
  1. Write out and balance each of the following equations (given in the Introduction).
  2. If iron(III) sulfate were formed, what mass of copper would be expected?
  3. If iron(II) sulfate were formed, what mass of copper would be expected?

Materials

Apparatus
Safety goggles
Lab apron
Two 250-mL beakers
Ring stand
Ring
Wire Gauze
Bunsen burner
Balance
Stirring rod
Matches

Reagents
Copper(II) sulfate (anhydrous)
Water

Safety

1. Safety goggles and a lab apron must be worn at all times in the laboratory.
2. If you come in contact with any solution, wash the contacted area thoroughly.
3. You will work with a flame in this lab. Tie back hair and loose clothing.
4. Do not drop matches into the sink. Dispose of burned matches in the trashcan after they are cool.

Procedure

1. Place 7.00 g of copper(II) sulfate in a beaker.

2. Add about 50 mL of water to the beaker.

3. Arrange the beaker and ring stand as shown in **Figure 1**.

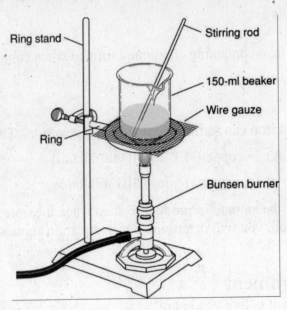

Ring stand

Stirring rod

150-ml beaker

Wire gauze

Ring

Bunsen burner

Figure 1
Apparatus to heat a solution

4. Light the Bunsen burner and place it under the beaker. Adjust the burner so the hottest part of the flame touches the bottom of the beaker.

5. Carefully heat and stir the mixture in the beaker. The solution should be hot, but not boiling. After all of the crystals have dissolved, remove the beaker from the heat.

6. Add 2.00 g of iron filings slowly to the hot copper sulfate solution while stirring. Record observations.

7. Allow the beaker to cool for 10-15 minutes.

8. Pour off (decant) the solution into a different beaker. Pouring the solution down a stirring rod is recommended. Make sure not to disturb the copper in the beaker.

9. Add a small amount of water (10 to 15 mL) to the copper and stir.

10. Let the copper settle to the bottom of the beaker and decant again.

11. Dry the copper as your teacher directs and determine its mass. Record this mass.

Cleaning Up

1. Place the copper in the waste container provided by your teacher.
2. Place the remaining solution in the labeled container.
3. Wash all glassware and return all materials to the proper locations.
4. Wash your hands thoroughly before leaving the laboratory.

Data/Observations

Complete the Data/Observations section for this experiment either on your Report Sheet or in your lab report as directed by your teacher.

1. Table 7.1 lists clues that a chemical reaction has occurred. Which of the observations you made support that a chemical reaction has occurred in the experiment?

2. What mass of copper was formed?

Analysis and Conclusions

Complete the **Analysis and Conclusions** section for this experiment either on your Report Sheet or in your lab report as directed by your teacher.

1. Which reactant was limiting? What observations support this? Also, use the chemical equation for the reaction that occurred in this lab and stoichiometric calculations to support your answer.

2. Why were the amounts chosen so this reactant (see number 1 above) was limiting? What would be a problem with having the other reactant as the limiting reactant?

3. Why was the copper washed with water (step 9 of the procedure)?

4. Why didn't the water added to the copper (II) sulfate have to be measured exactly?

5. What is the formula for the iron-containing compound that is formed when copper (II) sulfate and iron react? Support your answer.

Something Extra

Remove the copper metal and let the solution stand for a few days. What happens? Explain.

Experiment 38

Calorimetry and Limiting Reactants

Problem
How can we use the temperature change for a reaction to help us to identify an unknown acid?

Introduction
Recall from *World of Chemistry* Chapter 8 that the net ionic equation for the reaction of any strong acid with hydroxide is:

$$H^+(aq) + OH^-(aq) \rightarrow H_2O(l).$$

You may also recall that it is the formation of water from the hydrogen and hydroxide ions that provides a driving force for these reactions. This reaction releases a significant quantity of heat (the process is exothermic); the actual amount of heat depends of the number of moles of reactants consumed. For example, if we double the amount of each reactant, the amount of heat released will double.

You will be provided with solutions of sodium hydroxide, $NaOH(aq)$, and either sulfuric acid, $H_2SO_4(aq)$, or hydrochloric acid, $HCl(aq)$. Each of these solutions contains 1.0 mol of solute per 1.0 L of solution. The sodium hydroxide will be labeled; but the acid will not; they will simply be identified as **ACID A** and **ACID B**. It is your task to decide which one you have. To accomplish your goal, you will mix different volumes of the acid with a fixed volume of base, and then use the change in temperature to measure the extent to which the reaction has taken place.

You will carry out the neutralization reactions in a calorimeter and you will monitor the temperature of the reactions using the thermistor probe on your CBL unit. Differences in the quantity of energy evolved as heat, as shown by differences in temperature, will serve as the key to answering the question stated in the Problem. The Prelaboratory Questions will help you see how the data you obtain will help you solve the problem. Your teacher will tell you whether you are to use Acid A or Acid B.

Prelaboratory Assignment
✓ Read the **Introduction** and **Procedure** before you begin.
✓ Answer the Prelaboratory Questions.

1. Determine the number of moles of acid and base for each trial and record your results in a Summary Table (see **Analysis and Conclusions** for how to prepare this Table if your are not using a Report Sheet).

2. If two separate 50.0 mL samples of the NaOH are mixed with 25.0 mL portions of HCl and H_2SO_4, respectively, which combination should produce the greater quantity of heat energy? Why?

Materials

Apparatus
CBL2, LabPro, or similar interface
Temperature probe
Interface-appropriate software: (DataMate for CBL2 or LabPro)
Foam polystyrene cups (3)
Beaker, 250-mL, to hold calorimeter
Graduated cylinders, 25-mL (3)
Safety goggles
Lab apron

Reagents
Sodium hydroxide, NaOH(aq)
Acid A or Acid B

Safety

1. All three reagents are corrosive to skin and clothing. Handle with care and clean up all spills with large amounts of water. In the event of an acid spill on the lab bench, first neutralize the spill with baking soda.
2. If any reagent spills on your skin or clothing, flood the affected area with water.
3. Safety goggles and a lab apron must be worn at all times when working in the laboratory.

Procedure

Note: Throughout the description that follows, any reference to "CBL" is to be understood to mean CBL, CBL2, or LabPro.

Calorimeter Apparatus

The calorimeter used in this experiment is made of three white foam polystyrene coffee cups. Foam polystyrene, as you know from experience, is an excellent insulator (see **Figure 1**).

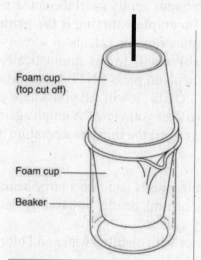

The double foam cup is placed in a 250-mL beaker, for stability. A temperature probe or standard thermometer is inserted through the top of the foam lid, as shown by the black line..

Foam cup (top cut off)

Foam cup

Beaker

Figure 1: The Calorimeter

Nest two Styrofoam cups and place them inside a 250-mL beaker. Cut away the top 1-inch from a third cup and pierce the bottom of the cut cup to insert a thermometer or digital probe. The probe or thermometer should touch (but not penetrate) the bottom of the inner cup. The sharp point on a digital thermometer can be used to pierce the upper cup for a digital thermometer or temperature probe. Use a cork borer to make a tight-fitting hole for a mercury thermometer.

If you are not using a Report Sheet for this experiment design a Data Table with the following headings. Indicate whether you have been assigned to test Acid A or Acid B.

Combination Tested	Initial Temperature	Final Temperature	ΔT

Note: Each trial will use a combination of 20.0 mL of sodium hydroxide, with a different volume of acid each time. The volumes of acid to be used are: 5.0 mL, 10.0 mL, 15.0 mL, 20.0 mL, and 25.0 mL.

1. Set up the interface to take readings every second for a period of 120 seconds. If you are called on to set the temperature range, select 0°C to 50°C. Do not start collecting data yet.

2. Measure 20.0 mL of sodium hydroxide into one 25-mL graduate and 5.0 mL of Acid A or B into the 25-mL graduate designated for use with that acid. See the Summary Table for the volumes to be used.

3. With the lid off, place the 20.0 mL sample of NaOH solution in the calorimeter. Briefly replace the lid, so that the probe reaches down into the sodium hydroxide. Be sure that the CBL is turned on and that the temperature of the sodium hydroxide appears on the screen. When all is ready, proceed to step 4. You will have to act quickly to ensure that you record all the data you need.

4. START the data collection. (With LabPro and CBL2, START is option #2 on the MAIN MENU.) Allow the CBL to record the sodium hydroxide temperature a few times, then remove the lid as you **quickly** add 5.0 mL of your acid. Be sure to add **all** of the acid. Carefully replace the lid, returning the probe to the solution.

   Holding the apparatus with both hands gently swirl the entire assembly to help the acid and base mix as effectively as possible. **Incomplete mixing is the primary cause for erratic data.** At the conclusion of the two-minute run, the calculator screen will re-plot the data with adjusted temperature scale. CBL2 and LabPro will do this automatically. (With CBL you must press ENTER to get the new plot.) When you press ENTER, DataMate will return you to the MAIN SCREEN. If you select option 6:QUIT, it will tell you where your data is stored (probably lists L1 and L2). Pressing ENTER will let you view the graph again. Use the left and right arrows to trace along the graph. Note and record the initial temperature and the highest temperature reached.

5. Pour the contents of the inner calorimeter cup into a large beaker (400-mL or larger) for later treatment. The contents may be neutral, acidic, or basic, so neutralization will be needed to render them safe for the sanitary system.
   Rinse the inside of the calorimeter with distilled water and blot it dry, lightly and gently with paper towel.

6. Repeat steps 3-5 for the remaining four volumes of acid. Remember that the volumes to be used are 5.0 mL, 10.0 mL, 15.0 mL, 20.0 mL, and 25.0 mL.

7. **(Optional.)** With your teacher's permission, repeat the entire experiment, this time using the other acid. The volumes of acid are the same for both.

Cleaning Up

1. Disassemble the CBL apparatus, taking care to thoroughly rinse the temperature probe with distilled water. Return all parts to their proper location, as your teacher directs.
2. Rinse the inner cup and lid of the calorimeter, then either leave them for the next class or dispose of them in an appropriate manner, as your teacher directs. The beaker is to be returned to its proper location.
3. Wash your hands before leaving the laboratory.

Analysis and Conclusions

Complete the **Analysis and Conclusions** section for this experiment either on your Report Sheet or in your lab notebook, as directed by your teacher. If you are not using a Report Sheet prepare a summary Table with the following headings:

Volume Acid A (or B)	Volume NaOH	Moles Acid A (or B)	Moles NaOH	Which is Limiting?	Temperature Change

1. Compare the temperature changes. Which combination produced the greatest change in temperature? Was your acid hydrochloric acid or sulfuric acid? Cite experimental evidence to support your choice.

2. In all but one of the trials there was a limiting reactant.
   a. Based on your answer to question 1, which cases involved a limiting reactant? Identify which reactant (acid or NaOH) was limiting and cite experimental evidence or mathematical proof for each case in which there was a limiting reactant.
   b. In which trial was there not a limiting reactant? Defend your choice.

3. Construct a graph of temperature change versus volume of acid added for your acid. On the same set of axes, but in a different color, sketch the graph you would expect if you were to use the other acid. Explain your reasoning.

4. When you did each trial, the calculator screen showed you a plot of temperature as a function of time. Sketch the shape of one of those graphs and account for what you see. Indicate on your sketch the initial and final temperatures.

Something Extra

1. Is the heat that is produced in the reaction between an acid and a base dependent on the strength of the acid or base?

   To answer this question measure the heat produced by the following combinations of acids and bases:
   a. Strong acid-strong base: HCl and NaOH.
   b. Strong acid-weak base: HCl and $NH_3(aq)$.
   c. Weak acid-strong base: $HC_2H_3O_2$ and NaOH.
   d. Weak acid-weak base: $HC_2H_3O_2$ and NH_3.

   Use 20.0 mL of each reactant.

   Note: Ammonia has a strong, unpleasant odor. Do these experiments in a fume hood or in a well-ventilated room.

2. Use a computer program such as Logger Pro™ to plot the results of each reaction in Something Extra, #1. Compare the plots in terms of ΔT values and the time needed to reach the maximum temperature in each case. Account for any differences you find.

Experiment 39

Synthesis of Manganese(II) Chloride

Problem
What special techniques are needed to synthesize a pure sample of a metal halide?

Introduction
In this experiment you will synthesize manganese(II) chloride by allowing a known mass of manganese metal to react with an excess of hydrochloric acid. From the theoretical yield (calculated from the mass of metal used) and the actual mass of product recovered, you will calculate the percentage yield. The balanced equation for the reaction is:

$$Mn(s) + 2\ HCl(aq)\ \rightarrow\ MnCl_2(aq) + H_2(g)$$

According to the equation, one mole (54.94 grams) of manganese will generate precisely one mole (125.84 grams) of manganese(II) chloride. Regardless of the actual mass of manganese used, the masses of reactant and product must be in this same ratio. Once the synthesis is complete and you have isolated your product, you will determine your actual yield by weighing the flask and product. You will then compare it with the expected yield (called the *theoretical yield*), by applying the ratio of the molar masses of the product and metal. From these two, you will calculate your percentage yield. As part of your error analysis, you will try to account for any deviation between the actual and theoretical yields.

Evaporation of the solution remaining in the flask after reaction will yield solid manganese(II) chloride. The product is very hygroscopic (water absorbing), so it will be necessary for it to cool in a location that will not allow absorption of atmospheric moisture. Such dry-atmosphere cooling chambers are called *desiccators*.

Prelaboratory Assignment
✓ Read the **Introduction** and **Procedure** before you begin.
✓ Answer the Prelaboratory Questions.

1. Explain the difference between *theoretical yield* and *actual yield*.

2. Why is it important for the manganese to react entirely before proceeding at step 5?

3. What is a desiccator, and why is it needed in this experiment?

4. **a.** If 10 M HCl(aq) accidentally spills or splashes on your skin or clothing, describe in detail what you should do.

   b. What, if anything, would you do differently if the acid spilled on the surface of the lab bench or hood?

5. When you are done with the manganese(II) chloride, how should you dispose of the product?

Materials

Apparatus
Milligram balance
Erlenmeyer flask, 10-mL
Sand bath
Tongs or forceps for handling (hot flask)
Desiccator
Safety goggles
Lab apron

Reagents
Manganese metal, 0.1 gram
10 *M* hydrochloric acid, 1.5 mL
Boiling stone

Safety

1. Always wear your safety goggles and a lab apron in the laboratory.
2. 10 *M* Hydrochloric Acid is severely corrosive to both skin and clothing. Use it only in the fume hood.
3. **The pipet used to dispense 10 *M* HCl is not to leave the hood area,** nor is it to be placed on the bench top or handed directly to the next user; either of these could result in acid burns, or contamination of reagents, or both.
4. Heating will be done on a sand bath; **the sand bath will be quite hot** (>200° C), but will not appear so.
5. **Contact Lens Alert:** Vapors containing hydrochloric acid are given off during the evaporation step.
6. The manganese may be contaminated with manganese sulfide, which will generate hydrogen sulfide gas, $H_2S(g)$. If you detect a rotten-egg odor, immediately remove your flask to the fume hood.
7. Despite the presence of the boiling stone, there may be some spattering during the evaporation. Keep your hands and face away from the flask, especially during the latter stages.

Procedure

1. Check that the sand bath is on and the controller is set at 40% power.

2. Add a small boiling stone to a labeled 10-mL Erlenmeyer flask and determine the mass of the flask and stone. Place about 0.05-0.1 g of manganese metal in the flask; then reweigh the flask with the boiling stone and the metal. Record both masses in your Data Table; be sure they are recorded to the full precision of the balance.

3. Take your flask to the fume hood and add about 1.5 mL (estimated) of 10 *M* hydrochloric acid, HCl(*aq*). *See Safety Note #5, above.*

4. Allow the acid to react with the metal until the sample has dissolved completely (about 5 to 15 minutes). All the manganese must be converted to manganese(II) chloride before you begin the evaporation. During the dissolving process, try to determine whether the reaction is exothermic or endothermic.

5. Place the flask on the hot sand bath to evaporate the water and unreacted HCl as shown in Figure 1. Continue until the water has entirely evaporated, leaving only a pinkish-white solid residue. If droplets persist on the flask walls, use tongs to lay the flask carefully on an angle, with the lip of the flask on the rim of the sand bath.

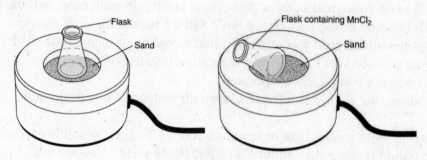

Figure 1
Evaporation in a sand bath

6. When the flask is thoroughly dry, use tongs or a hot-pad to remove it from the sand bath to a place where it can cool for 5-10 minutes. A desiccator made from a large coffee can with a drying agent in the bottom is ideal. The drying agent keeps the product from absorbing moisture from the air.

7. After the flask has cooled to room temperature, determine the mass of the flask and product.

8. Return the flask to the hot sand bath for an additional 3-5 minutes heating. Cool the flask in the desiccator, then measure its mass and record it in your Data Table. Ideally, this mass should agree to within 0.003g of the previous weighing.

Cleaning Up
1. Hand your labeled flask in to your teacher.
2. Manganese compounds require special handling for disposal; do not wash them down the sink.
3. Before leaving the laboratory, clean up all materials and wash your hands thoroughly.

Analysis and Conclusions
Complete the **Analysis and Conclusions** section for this experiment either on your Report Sheet or in your lab report as directed by your teacher

1. Calculate the mass and the number of moles of manganese metal used.

2. Use the final mass of the flask and product to find the mass and number of moles of manganese(II) chloride produced. This is your actual yield.

3. Calculate the mass of manganese(II) chloride you would expect to produce from the original mass of manganese metal. This is your theoretical yield.

4. Determine the percent yield of manganese(II) chloride.

$$100\% = \frac{\text{Actual Yield}}{\text{Theoretical Yield}} = \text{Percent Yield}$$

5. Predict the effect on your yield of these procedural errors. In each case, will the reported mass of manganese(II) chloride be too high or too low? Give a reason for each answer.
   a. The manganese metal used was impure; it had a coating of manganese oxide.
   b. Some of the solution was lost during evaporation because of spattering.
   c. The water was not completely evaporated.
   d. During cooling, the flask was left in the open air instead of in a desiccator.

6. Was your percent yield greater than or less than 100%? Suggest possible explanations for any significant variation (greater than about 5%) from 100% yield. You are not limited to the choices given in question 5. Base your answer on your own experience doing the experiment.

7. Explain how you arrived at the number of significant figures you show for your percent yield.

Experiment 40

The Energy Value of Nuts

Problem

How much heat is available from the combustion of peanuts, almonds or cashews? Can this heat be measured?

Introduction

When organic substances, such as nuts or other typical foodstuffs, are combusted in the presence of oxygen in the air, the byproducts of the combustion include carbon dioxide, water vapor, carbon (in the form of soot, which indicates incomplete combustion). Perhaps most important is the energy, released as heat. All of us depend upon food as a fuel, and when we metabolize food, energy is always released. Many people monitor their calorie intake, which is the amount of energy that is available in the foods they eat. In this experiment, various kinds of nuts are ignited with a match and allowed to burn as completely as possible. The energy released as heat is absorbed by a volume of water in a small beaker or flask. At the conclusion of the experiment, you will be able to compare the energy values of various nuts.

Prelaboratory Assignment

✓ Read the **Introduction** and **Procedure** before you begin.
✓ Answer the Prelaboratory Questions.

1. Paraffin wax has the chemical formula $C_{25}H_{52}$. Write a balanced equation for the combustion of paraffin in air.

2. How much energy is required to warm 100 g of H_2O from 20°C to 80°C if it is initially in a 140 g glass flask? The specific heat of water is 4.18 J/g °C; the specific heat of glass is 0.836 J/g °C.

3. Predict which of these three nuts – peanuts, cashews, almonds – will furnish the greatest amount of energy per gram. Explain briefly your hypothesis.

Materials

Apparatus
Calorimetry apparatus
150-mL beaker or Erlenmeyer flask
100-mL graduated cylinder
Thermometer
Centigram or milligram balance
Safety goggles
Disposable gloves
Matches
Lab apron

Reagents
Three samples of nuts
(peanuts, cashews, almonds,
pecans, etc.)

Safety

1. Wear safety goggles and lab aprons at all times in the laboratory.
2. No food or drink is allowed in the laboratory at any time.
3. Gloves may be worn to protect your hands from the charred nut remnants.
4. Be careful to position the beaker as securely as possible atop the apparatus.
5. The nuts may flame up quickly. Be sure the apparatus is a safe distance away from you.

Procedure

1. Obtain a calorimetry apparatus.

2. Select a nut. Be sure that a small hole has been bored into it. Determine the mass of the nut and record it.

3. Obtain a clean, dry 150-mL beaker or Erlenmeyer flask. Determine its mass and record it.

4. Place 70-80 mL of tap water into the beaker or flask. Measure the volume of water you use to the nearest 0.1 mL and record it. Determine the initial temperature of the water and record it.

5. Insert the skewer into the nut and suspend it in the calorimeter as shown in **Figure 1**.

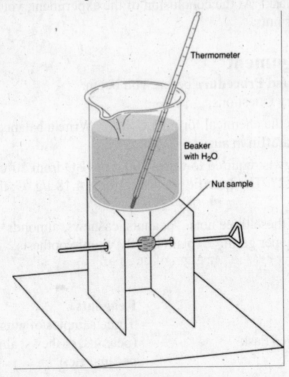

Figure 1

Apparatus for determining the energy in a nut

6. Set the beaker of water on top of the calorimeter platform as shown in **Figure 1**. Using a match, ignite the nut, and let it burn until the flame goes out.

7. Determine the highest temperature of the water. Record your value.

8. Carefully remove the charred nut remnant, determine and record its mass. **Note:** You may wish to wear disposable gloves to do this.

9. Repeat the experiment two more times with different nuts. Be sure to use fresh samples of water each time.

Cleaning Up

1. Dispose of the charred remnants as directed by your teacher.
2. Wipe off the calorimetry apparatus and return it to its proper location. Take care not to get the soot on yourself or on the table.
3. Clean all glassware and return to its proper location.
4. Wash your hands thoroughly before leaving the laboratory.

Analysis and Conclusions

Complete the **Analysis and Conclusions** section for this experiment either on your Report Sheet or in your lab report as directed by your teacher.

Complete the following analysis for each sample nut. Summarize your results in a Summary Table.
1. Determine the change in mass of the nut as a result of combustion.

2. Determine the change in temperature of the water and the beaker as a result of combustion of the nut.

3. Determine the heat absorbed by the water, using the equation $\Delta H = s \times m \times \Delta T$. The specific heat, s, for water is 4.18 J/g °C.

4. Determine the heat absorbed by the beaker. The specific heat for glass is 0.836 J/g °C.

5. Determine the total heat absorbed by the water and the beaker. **Note:** This heat is equal to the heat released by the nut.

6. Determine the total heat released per gram of nut.

Something Extra

1. A serving size of nuts or chips is often listed on food packages as one ounce (28.35 g). Using your lab data, determine the energy value per serving for each of the nut samples you used.

2. Critique the procedure used in this experiment. Do you expect the procedure to give an accurate energy value for the nuts? Explain your answer.

3. Compare your results to the nutritional information listed on the packages of commercially available products.

Experiment 41

Specific Heat of a Metal

Problem
How can the techniques of calorimetry be used to determine the specific heat capacity of a typical metal?

Introduction
When we wish to determine the amount of heat gained or lost during a process, we use a *calorimeter* (literally, a calorie measurer) in which a thermometer or temperature probe measures the changes in temperature that result from the gain or loss of thermal energy. The calorimeter is an insulated container that allows as little heat as possible to be lost to or absorbed from the surroundings. We might also expect the container and the thermometer to absorb or release some heat energy; this amount of heat is the *heat capacity* of the calorimeter. In highly accurate calorimetric experiments the heat capacity of the calorimeter is determined first, then that value is used in calculating the heat evolved in all subsequent experiments done with that calorimeter.

Knowledge of the magnitude of the temperature change that takes place in some process or reaction, the heat-absorbing capacity of the calorimeter, and the amount of one of the substances that is used in the process allows you to calculate the energy released or absorbed in terms of joules (or kilojoules) per mole of that substance. If all the heat released is used to heat water, the amount of heat involved can be found from the equation

$$q = (m_{water})(\Delta T)(1 \text{ cal/g°C})$$

where the last term, 1 cal/g K, is called the *specific heat (or the specific heat capacity) of water* and is, in fact, the definition of a calorie. One calorie is the amount of heat required to heat 1 g of water by 1 °C. The value for q can be changed to joules by the conversion factor, 1 calorie = 4.184 joules; thus, the equation which we will use is:

$$q = (m_{water})(\Delta T)(4.184 \text{ J/g°C}) \qquad \textbf{(Equation 1)}$$

The changes in temperature can be followed in a number of ways, the most familiar being an ordinary mercury (or alcohol)-in-glass thermometer. A thermometer is an analog device in which the expansion of mercury or alcohol corresponds to changes in temperature. Your brain converts the length of the liquid column to a digital output when you read the position of the column on the thermometer scale.

In a digital thermometer a tiny diode near the tip of the probe, powered by a battery, produces an output voltage that changes with temperature. This voltage, like the liquid in the thermometer, is an analog signal, but it can be easily converted to a digital signal by a computer chip. The signal is then read on the dial of the thermometer.

Prelaboratory Assignment

✓ Read the **Introduction** and **Procedure** before you begin.
✓ Answer the Prelaboratory Questions.

1. Since the specific heat of water is given in units of joules per *gram* degree Celsius why do we measure the volume of water in the calorimeter instead of its mass?

2. A 22.50-g piece of an unknown metal is heated to 100.°C, then transferred quickly and without cooling into 100. mL of water at 20.0°C. The final temperature reached by the system is 26.9°C.
   a. Calculate the quantity of heat absorbed by the water. Show all work.
   b. Determine the quantity of heat lost by the piece of metal. Show all work.
   c. Calculate the specific heat of the metal in J/g°C. Show all work.

3. What would be the effect on the value of the specific heat capacity of water if all temperatures were measured in kelvins (K) rather than degrees Celsius (°C)? Explain.

Materials

Apparatus
Milligram balance
Foam polystyrene cups, 6 oz (3)
250-mL beaker (2)
Thermometer, mercury- or alcohol filled, or digital
   or CBL2 or LabPro with DataMate application and temperature probe
25-mL graduated cylinders (2)
Hot plate or heating apparatus
Safety goggles
Lab apron
Hot pad or mitt

Reagents
Tap water
Metal shot

Safety

1. Wear safety goggles and a lab apron at all times in the laboratory.
2. Handle the tube of hot metal shot carefully, using a hot pad.
3. The heating unit and beaker are hot. Be sure to use care when working with them.

Procedure

Calorimeter Apparatus

The calorimeter used in this experiment is made of three white foam polystyrene coffee cups. Foam polystyrene, as you know from experience, is an excellent insulator. Nest two Styrofoam cups and place them inside a 250-mL beaker. Cut away the top 1-inch from a third cup, and pierce the bottom of the cut cup to insert a thermometer or digital probe. The probe or thermometer should touch the bottom of the inner cup. The sharp point on the digital thermometer can be used to pierce the upper cup for the digital thermometer or temperature probe. Use a cork borer to make a tight-fitting hole for a mercury or alcohol thermometer (see **Figure 1**).

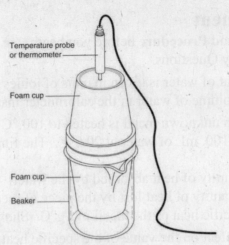

Temperature probe
or thermometer

Foam cup

Foam cup

Beaker

Figure 1
The calorimeter

1. Fill a 250-mL beaker about halfway with water. Place it on your hot plate or heating apparatus and begin heating the water until it reaches the boiling point.

2. Place 50.0-mL of water in the calorimeter. Note and record the temperature in a Data Table.

3. Obtain a tube containing about 10.0 g of metal shot. Note and record the identity and mass of shot (given on the label of the tube) in your Data Table. Place the tube of metal shot in the boiling water bath for at least 3 minutes. This is to ensure that the temperature of the shot is 100°C, the temperature of boiling water.
   Note: The tube of metal shot is hot. Use beaker tongs or protect your hand with a cloth or mitt in the following step.

4. Quickly and carefully transfer the metal sample at 100°C to the room temperature water in the calorimeter. Quickly place the lid containing the thermometer or temperature probe back on the calorimeter.

5. Note and record the highest temperature reached by the contents of the calorimeter.

6. If time permits, or if you teacher so directs, repeat the experiment, starting with fresh, cool water in the calorimeter, and a dry sample of metal shot. If the results of the second trial differ greatly from the results of the first, carry out a third trial.

Cleaning Up

1. Water can be disposed of down the drain. Pour off as much as you can without losing any of the metal shot.
2. Pour the metal (and any remaining water) out onto several thickness of paper towel. The metal should be returned to its proper location. Your instructor will direct you in what is to be done with the polystyrene cups.
3. Wash your hands thoroughly before leaving the laboratory.

Analysis and Conclusions

Complete the **Analysis and Conclusions** section for this experiment either on your Report Sheet or in your lab report as directed by your teacher.

1. Calculate the quantity of heat gained by the water, using Equation 1. If you carried out additional trials, report the results of all trials, as well as an average (mean) value. You need to show your work for only one of the trials.

2. Assume that the quantity of heat lost by the metal is equal to the quantity of heat gained by the water. Use Equation 1 to determine the specific heat, **s**, of the metal. Be sure you use ΔT for the metal in your calculation. As before, if you carried out multiple trials, report the result for each trial, as well as a mean value. Show the calculations for the first trial only.

3. Consider the assumption you were asked to make in **2**.
   a. Explain why the assumption is not entirely valid.
   b. Is using the assumption in **2** likely to give a value for the specific heat of the metal that is too high or too low? Explain.

4. Look up the value of the specific heat of your metal in the *Handbook of Chemistry and Physics*. Calculate your percent error, using the following equation (note the "absolute value" signs).

$$\% \text{ error} = \frac{|\text{(accepted value) - (experimental value)}|}{\text{(accepted value)}} \times 100\%$$

Something Extra

1. Any calorimeter absorbs a certain amount of the heat released. Knowing this, is your value of the specific heat of the metal more likely to be higher or lower than the accepted value? Explain.

2. Develop a procedure for determining the heat capacity of the calorimeter that you used. With your teacher's approval, carry out your procedure.

Experiment 42

Which is Your Metal?

Problem

What is the identity of the metal you are given?

Introduction

The specific heat capacity is the amount of energy required to change the temperature of one gram of a substance by one Celsius degree. Each substance has a unique specific heat capacity. In this lab you will use a determination of the specific heat capacity to identify an "unknown" metal.

Specific Heat Capacities (J/g°C)

Water: 4.183

Zn: 0.3882

Sn: 0.2271

Pb: 0.1280

Al: 0.8910

Cu: 0.3844

Prelaboratory Assignment

✓ Read the entire experiment before you begin.

✓ Answer the Prelaboratory Questions.

1. How do you know the final temperature of the metal?

2. Why do you use a hot water bath to heat the metal?

3. Why do you measure the volume of the water in the calorimeter when we need to know the mass of the water for the calculations?

Materials

Apparatus

Safety goggles

Lab apron

2 foam cups with lids (2 holes)

Ring Stand and ring

Wire gauze

Matches

Graduated cylinder

Bunsen burner

Tongs

Stirrer

Thermometer

Reagents

Water

"Unknown" metal

Safety

1. Thermometers are fragile. Be careful in handling them, and never use a thermometer as a stirring rod. If you are using a mercury thermometer and it breaks, notify your teacher immediately. Mercury vapors are poisonous.
2. Safety goggles and a lab apron must be worn at all times in the laboratory.
3. You will work with a flame in this lab. Tie back hair and loose clothing.
4. Do not drop matches into the sink. Dispose of burned matches in the trashcan after they are cool.

Procedure

1. Find the mass of your metal sample. Record this value.

2. Place one foam cup into the other. This is your calorimeter.

3. Measure 75.0 mL water and place it in your calorimeter. Record the temperature of the water.

4. Set up the ring and ring stand and place the 250-mL beaker on the wire gauze as shown in **Figure 1.**

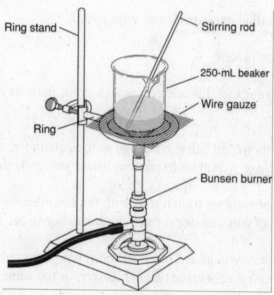

Ring stand

Stirring rod

250-mL beaker

Wire gauze

Ring

Bunsen burner

Figure 1
Apparatus for boiling water

5. Add about 100 mL of water to the beaker and place the metal in the water.

6. Use a Bunsen burner to heat the water to boiling. Allow the metal to remain in the boiling water for at least three minutes. Record the temperature of the boiling water.

7. Using the tongs, quickly transfer the metal from the boiling water to the calorimeter. Cover the calorimeter.

8. Gently stir the water in the calorimeter for several seconds (do not use the thermometer).

9. Record the highest temperature reached.

10. Repeat these steps two more times.

Cleaning Up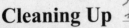

1. Return your metal sample and foam cups to your teacher.
2. Return all materials to their proper locations.
3. Wash your hands thoroughly before leaving the laboratory.

Analysis and Conclusions

Complete the **Analysis and Conclusions** section for this experiment either on your Report Sheet or in your lab report as directed by your teacher.

1. Calculate the heat transferred to the water in the calorimeter in each trial.

2. How does the heat transferred to the water in the calorimeter compare to the heat transferred from the metal in the calorimeter?

3. Calculate the specific heat capacity of your metal in each trial.

4. Calculate the average specific heat capacity of your metal.

5. What is the identity of your metal?

6. Assuming the identity of your metal is correct, what is your percent error in the specific heat capacity?

7. Both of the following errors would cause a change in the calculated specific heat capacity for your metal. Tell if the change would be to raise or lower your calculated value of the specific heat capacity. Explain.
   a. A significant amount of water is transferred with the hot metal.
   b. The metal "cools off" as you transfer it from the hot water to the calorimeter.

8. Suppose that in the procedure you added metal at room temperature to hot water. How do you think this would affect your percent error (higher, lower, or the same)? Explain your reasoning.

Something Extra

Will you get similar results if you use a beaker instead of nested foam cups? What about a paper cup? Try it.

Experiment 43

Stoichiometry and Calorimetry

Problem
How is it possible to use calorimetry to determine the stoichiometric ratio for a reaction?

Introduction
In this experiment, you will determine the enthalpy change for a reaction, and at the same time study the stoichiometry of the reaction. There are two goals in this experiment. The first is to establish the stoichiometry for the reaction between sodium sulfite and household bleach, an aqueous solution of sodium hypochlorite. The second is to determine the heat of the reaction. To accomplish both goals, you will mix the reactants in a series of volume ratios and measure the temperature change in each case. For each trial you will use a total of 24.0 mL of combined solutions. Since both reactant solutions are 0.50 M, the ratio of volumes will be the same as the mole ratio for that trial. Thus, a trial, in which 6.0 mL of sodium sulfite is added to 18.0 mL of dilute bleach would have a 3:1 molar ratio of hypochlorite ion to sulfite ion.

The trial for which you observe the largest temperature change must be the one with the correct mol ratio of the reactants. From the ΔT for this "best" combination and knowledge of the number of moles of each reactant present you will be able to calculate heat for the reaction.

Prelaboratory Assignment
✓ Read the **Introduction** and **Procedure** before you begin.
✓ Answer the Prelaboratory Questions.

1. Hypochlorite is an oxidizing reagent. In this reaction the sulfite ion is oxidized to sulfate and the chlorine atom in the OCl⁻ ion will either become molecular chlorine or chloride ion.

   a. In addition to the method used in this experiment, how else could you decide whether the chlorine atom in hypochlorite becomes chloride ion or molecular chlorine?

   b. Write balanced molecular equations for the formation of each of the two possible chlorine products. Remember these are alkaline solutions, so hydroxide ions (from sodium hydroxide) may also be involved.

   c. How will you decide which of the two equations correctly represents the reaction between sulfite and hypochlorite?

Materials

Apparatus
50-mL beaker
100-mL beaker
250-mL beaker
Paper towels
Foam polystyrene cups, 6-oz (1)
25-mL graduated cylinders (2)
Thermometer or CBL-interfaced probe
Safety goggles
Lab apron

Reagents
Solutions of:
    sodium hypochlorite, NaOCl
    sodium sulfite, Na_2SO_3
Distilled water (wash bottle)

Safety

1. Wear safety goggles and a lab apron whenever you are working in the laboratory.
2. Sodium hypochlorite is a skin irritant. Avoid contact with skin.

Procedure

1. Prepare the calorimeter set-up as illustrated in **Figure 1**. The system consists of a 50-mL beaker, wrapped in paper towel for insulation, then inserted into a 100-mL beaker. The lid is an inverted 6-oz expanded polystyrene cup; the probe or thermometer is inserted through the base of the cup, as shown in **Figure 1**.

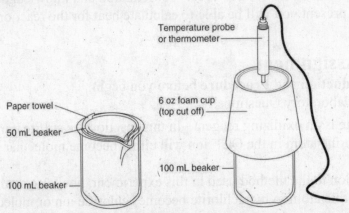

Figure 1
Apparatus for measuring temperature change during the reaction

Notes
✓ If a computer- or CBL-interfaced probe is used, it may be necessary to use a larger cup for the lid, trimmed to fit more closely to the top of the beakers. Most interfaced thermistor probes have bushings that can be used to make a good seal with the cup.
✓ If a standard glass thermometer is used, drill a small hole in the cup using a #2 cork borer. Do this very carefully, since a snug fit is essential for good results.
✓ A digital thermometer generally has a pointed tip which can be used to make its own opening. This opening will also fit a thermistor probe well.
✓ Regardless of the type of temperature sensor used, once it is in place it should not be removed so that the opening is not widened, thus reducing heat-retention.

2. Label separate 25-mL graduated cylinders for the two reactant solutions, Na_2SO_3 and $NaOCl$. Keep these two separate to avoid contamination.

3. Each trial follows the same basic sequence:
   a. Always place the reactant solution for which the volume is larger in the calorimeter first. It is necessary do this so that the thermometer or temperature probe is completely submerged.
   b. The lid containing the thermometer or probe is placed on the calorimeter and the temperature of the solution is recorded.
   c. The lid is removed and the second reagent is added *all at once*. The lid, containing the thermometer or probe, is immediately replaced and the calorimeter system is gently swirled to ensure complete mixing.
      ✓ The highest temperature reached by the system is noted and recorded
      ✓ The lid is removed and the contents of the system are rinsed down the drain.
      ✓ The inner beaker of the calorimeter is rinsed with distilled water and dried, in preparation for the next trial.

4. If you are using an electronic interface, set it up to read the temperature at 1-second intervals for a total of 3 minutes per run (you may find that two minutes is long enough; if so, make the appropriate changes).

Mixing Volumes

Trial	sodium hypochlorite	sodium sulfite	
1	18.0 mL	6.0 mL	
2	16.0	8.0	
3	14.4	9.6	**Note:** 14.5: 9.5 is close enough
4	12.0	12.0	
5	9.6	14.4	
6	8.0	16.0	
7	6.0	18.0	

Cleaning Up

1. All solutions remaining after the reaction may safely be poured down the drain.
2. Clean all glassware and return it to its proper location.
3. Your teacher will instruct you about what to do with the foam cups and CBL apparatus.
4. Wash your hands thoroughly before leaving the laboratory.

Analysis and Conclusions

Complete the **Analysis and Conclusions** section for this experiment either on your Report Sheet or in your lab report as directed by your teacher.

If you are not using a Report Sheet for this experiment design a Summary table with the following headings:

Trial	Mol OCl⁻/mol SO₃²⁻	q (J)	ΔH (kJ/mol NaOCl)

1. For each trial calculate **q**, the quantity of heat released by the reaction and enter your result in the Summary Table. Assume that each of the solution combinations has the same specific heat as water, 4.184 J/g°C. **Note:** The total mass of solution being heated is 24.0 g in each case.

2. To determine the stoichiometry of this reaction we need to establish the mole ratio that produces the maximum amount of heat.
   a. Use the volume and concentration of sodium hypochlorite to calculate the number of moles of hypochlorite ion present in each reaction system.
   b. Use the numbers of moles just calculated and the value of **q** for each reaction (from **1**) to calculate values of ΔH for each reaction in kJ/mol NaOCl. Record these values in the Summary Table.

3. Based on your answer to **2b**, which combination produced the greatest amount of heat per mole of sodium hypochlorite consumed? This must be the reaction combination that is closest to the actual stoichiometry of the reaction. Choose the equation from the Prelaboratory Questions that matches the mole ratio of sodium hypochlorite to sodium sulfite for your experimental result.

Something Extra

1. You were told to base your decision on the number of moles of sodium hypochlorite consumed. That was an arbitrary choice. Do the calculations necessary to show that the same results would be obtained and the same decision would have been made had you based your calculations in **2b** on the moles of sodium sulfite consumed, instead of on moles of sodium hypochlorite.

2. Repeat the experiment, using the same concentrations and volume combinations, but replace the sodium sulfite with either:
   a. sodium thiosulfate, $Na_2S_2O_3$, or
   b. sodium iodide. Follow the instructions for the original experiment.

Experiment 44

Heat of Reaction

Problem

What is the relationship among the heats of reaction of several chemical reactions?

Introduction

In this experiment you will measure and compare the quantity of heat released in three chemical processes. Each reaction involves the preparation of a solution. Each preparation causes a change in temperature. You will calculate of the energy change for these reactions after you measure the temperature changes. You will make several assumptions to assist in the calculations:

1. Any small heat losses to the surroundings will be neglected.
2. The aqueous solutions that are prepared have the same specific heat capacity as water: 4.18 J/g°C.
3. The solutions are so dilute that their densities are the same as that of water (1 g/mL).

The processes that are to be compared in this experiment are:

1. Solid sodium hydroxide dissolves in water to form an aqueous solution of ions.
$$NaOH(s) \rightarrow Na^+(aq) + OH^-(aq) \qquad \Delta H_1 = - x_1 \text{ Joules}$$

2. Solid sodium hydroxide reacts with an aqueous solution of hydrogen chloride to form water and an aqueous solution of sodium chloride.
$$NaOH(s) + H^+(aq) + Cl^-(aq) \rightarrow H_2O(l) + Na^+(aq) + Cl^-(aq) \qquad \Delta H_2 = - x_2 \text{ Joules}$$

3. An aqueous solution of sodium hydroxide reacts with an aqueous solution of hydrogen chloride to form water and an aqueous solution of sodium chloride.
$$Na^+(aq) + OH^-(aq) + H^+(aq) + Cl^-(aq) \rightarrow H_2O(l) + Na^+(aq) + Cl^-(aq)$$
$$\Delta H_3 = - x_3 \text{ Joules}$$

Prelaboratory Assignment

✓ Read the **Introduction** and **Procedure** before you begin.
✓ Answer the Prelaboratory Questions.

1. Write net ionic equations for the three processes in this experiment.
2. Use Hess's Law to find the heat of reaction for C(graphite) → C(diamond) given the following information:

C(graphite) + $O_2(g)$ → $CO_2(g)$	ΔH = -392.2 kJ/mol
C(diamond) + $O_2(g)$ → $CO_2(g)$	ΔH = -393.7 kJ/mol

3. For any given chemical reaction, how does the heat of the reaction depend on the mass of substance used in the reaction.
4. How many Joules are required to warm 250 mL of water from 20.5°C to 43.7°C?

Materials

Apparatus
Foam polystyrene cups (3)
100-mL graduated cylinder
Thermometer
Centigram or milligram balance
Safety goggles
Lab apron
Disposable gloves (optional)

Reagents
Sodium hydroxide, solid pellets
Sodium hydroxide solution, 2.0 M
HCl solution, 1.0 M
HCl solution, 2.0 M
Household "white" vinegar (for cleanup)

Safety

1. Wear safety goggles and a lab apron at all times in the laboratory.
2. Gloves may be worn to protect your hands from the caustic chemicals.
3. Do not handle sodium hydroxide pellets with your fingers; the NaOH is caustic.
4. Be careful not to get any of the HCl solutions on your hands; though the solutions are dilute, HCl is a caustic material.

Procedure

Reaction 1

1. Obtain 3 foam cups, label them cup 1, cup 2, and cup 3.

2. Measure 100 mL of cool tap water with the graduated cylinder and place it into foam cup 1.

3. Wait several minutes until a constant temperature is reached. Measure this temperature as precisely as possible, and record it.

4. Obtain a sample of about 4 g of sodium hydroxide pellets. Determine the mass of your sample to the nearest 0.01 g and record it. **Note:** Do not handle sodium hydroxide pellets with your fingers.

5. Pour the solid NaOH into the water.

6. Place the thermometer into the solution and, while holding the thermometer swirl the foam cup gently but continuously until the sodium hydroxide is dissolved. Record the highest temperature reached.

7. Place the solution in the waste container provided and rinse the foam cup thoroughly with water.

Reaction 2

1. Measure 100 mL of 1.0 M HCl solution with the graduated cylinder and place it into the second labeled (cup 2).

2. Wait until a constant temperature is reached. Measure this temperature as precisely as possible, and record it.

3. Obtain another sample of about 4 g of sodium hydroxide pellets. Determine the mass of your sample to the nearest 0.01 g.

4. Pour the solid NaOH into the cup containing the 1.0 M HCl.

5. Place the thermometer into the solution and, while holding the thermometer swirl, the foam cup gently, but continuously, until the sodium hydroxide is dissolved. Record the highest temperature reached.

6. This solution is neutral; wash it down the drain with large amounts of water. Rinse the cup thoroughly with water. If it is to be used again, dry it inside and out.

Reaction 3

1. Measure 50.0 mL of 2.0 M HCl into foam cup 3 whose mass has been determined and recorded.

2. Into a separate container, measure 50.0 mL of 2.0 M NaOH.

3. Both of these solutions should be at, or slightly below, room temperature. Check this with the thermometer (rinse and dry the thermometer before changing from one solution to the other). Record the temperature of each solution.

4. Add the NaOH solution to the HCl solution in foam cup 3. Mix quickly and record the highest temperature reached.

5. Dispose of the solution down the drain with lots of water.

Cleaning Up

1. Wash all glassware and return it to its proper location.
2. Your teacher will tell you what to do with the foam cups.
3. Wash your hands thoroughly before leaving the laboratory.

Analysis and Conclusions

Complete the **Analysis and Conclusions** section for this experiment either on your Report Sheet or in your lab report as directed by your teacher.

Analysis Questions 1 – 5 are to be done for each of the three reactions. Show your calculations for Reaction 1, then record the results for it and the other two reactions in a Summary Table.

1. Determine the change in temperature for the reaction.

2. Determine the number of moles of sodium hydroxide used in the reaction.

3. Determine the heat absorbed by the solution (which raises its temperature).

4. Determine the total heat released by the reaction.

5. Determine the heat evolved per mole of NaOH for the reaction.

6. Calculate the sum of ΔH_1 and ΔH_3. Find the difference between that sum and ΔH_2.

7. Calculate the percent difference between $(\Delta H_1 + \Delta H_3)$ and ΔH_2, using the following formula:

$$\% \text{ difference} = \frac{|(\Delta H_1 + \Delta H_3) - \Delta H_2| \ (100\%)}{\Delta H_2}$$

8. Refer to the net ionic equations that you wrote in the Prelaboratory Assignment. Now, using Hess's Law, add the first and third equations together. Cancel any reaction species which appear on both sides of the equation. Compare the sum of those two equations to the second equation.

9. In Process 1, ΔH_1 represents the heat evolved as solid NaOH dissolves. Look at the net ionic equations for the second and third reactions. In a few words, make similar statements about what each of the heats of reactions represents.

10. Suppose you had used 8.00 g of NaOH in Reaction 1. How would this have affected the change in temperature? Predict the heat that would have been evolved in the reaction. What effect would increasing the mass of NaOH used have had on your calculation of ΔH_1? Explain.

Something Extra

1. Use chemical resources to learn some of the practical uses of NaOH.

2. What would have been the impact on your results if:
   a. the NaOH pellets that you used were contaminated with a nonreactive substance.
   b. you had mistakenly used 1.0 M HCl rather than 2.0 M HCl in Reaction 3.
   c. you failed to stir the solutions in all three reactions.

3. Sodium hydroxide and hydrogen chloride are examples of substances which dissolve exothermically in water. Use chemical resources to identify at least two substances which dissolve endothermically in water.

Experiment 45

Heats of Reaction and Hess's Law

Problem
What is Hess's Law, and how does it work?

Introduction
In 1840, G.H. Hess recognized that the energy change that accompanies a reaction depends only on what the reactants and products are, not in the way in which one is converted to the other. This came to be known as Hess's Law. It will be your goal in this experiment to demonstrate for yourself that this statement is a valid one. To do that, you will measure the energy change for two different reactions, then combine them to predict the energy change for another reaction. Here are the three reactions; notice that all species that ionize in water are shown as ions, while solids and non-ionizing substances are shown in molecular form. Also notice that the first "reaction" is really just the dissolving of solid sodium hydroxide in water.

(1) Dissolving sodium hydroxide in water:

$$NaOH(s) \rightarrow Na^+(aq) + OH^-(aq)$$

(2) The reaction between solid sodium hydroxide and acetic acid:

$$NaOH(s) + HC_2H_3O_2(aq) \rightarrow H_2O(l) + Na^+(aq) + C_2H_3O_2^-(aq)$$

(3) The reaction between aqueous solutions of sodium hydroxide and acetic acid:

$$Na^+(aq) + OH^-(aq) + HC_2H_3O_2(aq) \rightarrow H_2O(l) + Na^+(aq) + C_2H_3O_2^-(aq)$$

You will measure the heats of reactions for the first two, then combine them to predict the quantity of heat that will be transferred in the third reaction. You will then measure the heat transfer of the third reaction directly to see how close your prediction came. The method of combining the first two reactions is the subject of the Prelaboratory Questions. The apparatus may be familiar to you from other experiments.

Prelaboratory Assignment
✓ Read the **Introduction** and **Procedure** before you begin.
✓ Answer the Prelaboratory Questions.
  1. Rewrite equation (3) of the **Introduction** in net ionic form.
  2. Recopy equations (1) and (2) from the **Introduction**, but reverse equation (1), so that solid sodium hydroxide appears on the products side of the arrow and the dissolved ions appear as reactants. Now add the two equations just as you would in algebra class when solving two equations in two unknowns.

3. Show that the result you obtained by combining reversed equation (1) and equation (2) is identical to the net ionic equation for reaction (3).

4. All three reactions that you run in this experiment result in temperature increases. Are the reactions endothermic or exothermic? Explain.

5. You will use approximately 2.00 g of solid sodium hydroxide in two of the three reactions. Convert 2.00 g of NaOH to moles of NaOH.

Materials

Apparatus
CBL2, LabPro, or similar interface
Temperature probe (If the Direct Connect
    probe is used, a CBL-DIN adapter is
    also needed.)
Application: DataMate for CBL2 or LabPro
Foam polystyrene cups (3)
Beaker, 250-mL, to hold calorimeter
Beaker, 50-mL or similar, as a weighing container for NaOH
Graduated cylinder, 50- or 100-mL
Graduated cylinders, 25-mL (2)
Safety goggles
Lab apron

Reagents
Sodium hydroxide pellets, NaOH(s)
Sodium hydroxide, NaOH(aq), 2.0 M
Acetic acid, HC$_2$H$_3$O$_2$(aq), 1.0 M
Acetic acid, HC$_2$H$_3$O$_2$(aq), 2.0 M

Safety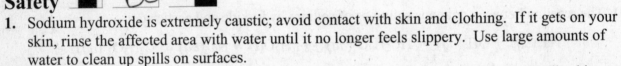

1. Sodium hydroxide is extremely caustic; avoid contact with skin and clothing. If it gets on your skin, rinse the affected area with water until it no longer feels slippery. Use large amounts of water to clean up spills on surfaces.
2. While acetic acid is considered a weak acid, it should be considered corrosive. Handle skin contact and spills as you would any acid spill.
3. Wear safety goggles and a lab apron at all times in the laboratory.

Note: **Throughout the description that follows, any reference to "CBL" is to be understood to mean CBL, CBL2, or LabPro.**

Procedure

1. Assemble the calorimeter as shown in **Figure 1**. Two cups are nested and placed in the beaker for stability. The third cup is trimmed about 1 cm below the rim and inverted as a lid. It should fit snugly into the opening of the other two cups. Use the tip of a digital thermometer (or similar instrument) to drill a hole in the lid that is just large enough to accept the temperature probe. The tighter the fit, the better the calorimeter will contain the heat that is released by the reaction.

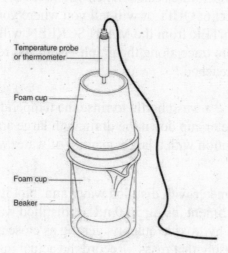

Temperature probe
or thermometer

Foam cup

Foam cup

Beaker

Figure 1: the Calorimeter

Once the temperature probe has been inserted into the calorimeter lid, avoid moving it any more than necessary, to maintain the snug fit.

2. Set up the interface to take readings every second for a period of 240 seconds. If you are called on to set the temperature range, select 0°C to 50°C. Do not start collecting data yet. **Note:** You will have to work quickly and efficiently while carrying out the reactions with solid sodium hydroxide. The NaOH pellets absorb water rapidly from the air. This will make them tend to stick to the weighing container.

3. Use the large graduated cylinder to place 50.0 mL of 1.0 M acetic acid, $HC_2H_3O_2(aq)$, into the calorimeter. Place the lid, containing the temperature probe, on the calorimeter.

4. Into your small beaker, weigh 2.00 g of NaOH pellets. **Note:** The pellets are of varying size, and you may not be able to get 2.00 grams exactly. Get as close as you can and record the actual mass of NaOH used, ± 0.01 g. Do not spend time trying to get 2.00 grams exactly.

5. START the data collection (with LabPro and CBL2, START is option #2 on the MAIN MENU). Allow the CBL to record the acetic acid temperature a few times, then remove the lid as you **quickly add all** of the sodium hydroxide pellets. Carefully replace the lid, returning the probe to the solution.

Holding the apparatus with both hands, gently swirl the entire assembly to help the pellets dissolve in the acid. At the conclusion of the four-minute run, the calculator screen will re-plot the data with the adjusted temperature scale. When you press ENTER it will return you to the MAIN SCREEN. If you select 6: QUIT, it will tell you where your data is stored (probably lists L1 and L2). Pressing ENTER from the MAIN SCREEN will let you view the graph again. Use the left and right arrows to trace along the graph. Note and record the initial temperature and the highest temperature reached.

6. Lift the calorimeter lid and use a wash bottle to rinse the temperature probe. Flush the contents of the inner calorimeter cup down the drain with large amounts of water. The contents are essentially neutral, so dilution with a large quantity of water will render them safe for the water system.

Rinse the inside of the calorimeter with distilled water and blot it dry, lightly and gently with paper towel. Repeat the experiment, using 50.0 mL of distilled water, in place of the acetic acid. Again, weigh out the sodium hydroxide quickly, getting as close as you can to 2.00 g, but do not waste time trying to get precisely that mass. Record the actual mass used.

Follow the procedure from step 4 to determine the quantity of heat released when solid sodium hydroxide dissolves in water. At the conclusion of the run, lift the calorimeter lid and use a wash bottle to rinse the temperature probe into the calorimeter cup.

In this case, the contents of the calorimeter are not neutral. Pour them into a 250-mL beaker, add a drop or two of phenolphthalein, then add acetic acid (or vinegar) until the pink color disappears. Then rinse the contents of the beaker down the drain, using a large volume of water. Rinse the calorimeter with distilled water and dry the inside of the calorimeter once more.

7. Repeat the experiment, using 25.0 mL each of 2.0 M sodium hydroxide and 2.0 M acetic acid. This time you can expect the temperature to rise more sharply, then begin to fall after about 2 minutes.

Cleaning Up

1. Rinse the temperature probe with distilled water, catching the rinsings in the calorimeter. The solution remaining in the calorimeter is essentially neutral and may be rinsed down the sink with water.
2. Your teacher will direct you regarding where the interface and temperature probe are to be placed. Be sure the temperature probe has been cleaned and dried before putting it away.
3. The calorimeter lids may be discarded in the waste basket unless your teacher directs you to save them for another class. The two other cups are to be rinsed thoroughly and blotted dry. They can be saved for another use.
4. Wash your hands thoroughly before leaving the laboratory.

Analysis and Conclusions

Complete the **Analysis and Conclusions** section for this experiment either on your Report Sheet or in your lab report as directed by your teacher.

1. Calculate q, the quantity of heat released by each of the reactions. Show your work. Be sure your answers are clearly identified and have the proper units.
   Reaction 1   Solid sodium hydroxide dissolves in water.
   Reaction 2   Solid sodium hydroxide reacts with 1.0 M acetic acid solution.
   Reaction 3   2.0 M sodium hydroxide reacts with 2.0 M acetic acid.

2. In the first two reactions, you used solid sodium hydroxide. Convert the actual masses of sodium hydroxide used to moles. Show your work.

3. The quantities and concentrations of the two solutions used in Reaction (**3**) were chosen so that 0.0500 moles of sodium hydroxide would be present. Use this information and the numbers of moles of NaOH you calculated in **2** to convert your values of q from question **1** to ΔH values in kJ/mol.

4. Recall that when you combined the equations for Reactions (**1**) and (**2**) in the second Prelaboratory Question, you reversed the direction of Reaction (**1**). If a reaction is exothermic in one direction, reversing the direction (products → reactants) changes the sign, but not the magnitude, of ΔH. Use your results from **question 3** to determine values for ΔH_1, ΔH_2, and ΔH_3, as identified below. Watch signs and be sure to include appropriate units.

$$Na^+(aq) + OH^-(aq) \rightarrow NaOH(s) \qquad\qquad \Delta H_1$$

$$NaOH(s) + HC_2H_3O_2(aq) \rightarrow H_2O(l) + Na^+(aq) + C_2H_3O_2^-(aq) \qquad \Delta H_2$$

$$Na^+(aq) + OH^-(aq) + HC_2H_3O_2(aq) \rightarrow H_2O(l) + Na^+(aq) + C_2H_3O_2^-(aq) \qquad \Delta H_3$$

5. Calculate the sum of ΔH_1 and ΔH_2. Show your work.

6. Because the sum of the first two equations of question 4 is the same as the third equation, the combined values of ΔH_1 and ΔH_2 should equal ΔH_3. Determine the percent deviation between the sum of ΔH_1 and ΔH_2, and the value you obtained for ΔH_3 by dividing the difference by the value for ΔH_3. Note the absolute value signs in the equation.

$$\frac{|\{\Delta H_1 + \Delta H_2\} - \{\Delta H_3\}|}{\Delta H_3} \times 100\% =$$

7. In this experiment some loss of heat energy is unavoidable. Discuss some of the ways in which heat energy may not have been fully accounted for in this experiment.

Experiment 46

Sunprint Paper Photography

Problem
What happens when light interacts with a chemically-treated piece of paper? Can light prompt a chemical reaction?

Introduction
In early investigations of the structure of atoms, a connection between light and chemical reactions was observed. Scientists became interested in reactions in which light was produced as well as in reactions which were initiated by light. High-energy ultraviolet light speeds up many chemical reactions. Sometimes this is useful (in the laboratory) and sometimes it is not desirable (such as in DNA mutation). In general when an atom or molecule absorbs light its electrons become excited (have higher energy) and become more reactive.

The cyanotype photographic process was developed by John Herschel in the 1840's. Amateur photographers began using cyanotype paper near the end of the 19th century because it was simple and inexpensive. Architects used this paper for copying drawings, but cyanotype paper was not used by serious photographers because of the bright color of the prints.

In this experiment you will prepare a photoactive solution which will be used to make sunprint paper. Then you will use the sunprint paper to "photograph" objects. Before coming to class you will need to find or make some objects to photograph. Be creative! Interesting or strange shapes work best. Some possibilities include keys, key rings, leaves, black and white negatives, paper cut-outs.

Prelaboratory Assignment
✓ Read the **Introduction** and **Procedure** before you begin.
✓ Answer the Prelaboratory Questions.

1. What determines the energy or color that is emitted when object glows or emits electromagnetic radiation?

2. Oxidation-reduction reactions are defined as processes in which electrons are transferred from one substance to another. Why is light often responsible for initiating oxidation-reduction reactions?

3. Rank the following types of electromagnetic radiation in increasing order of wavelength and in increasing order of energy: ultraviolet light, yellow light, infrared radiation, x-rays, blue light, radio waves.

Materials

Apparatus
Small beaker
Pie pan
Filter paper circles
Centigram or milligram balance
Graduated cylinder
Disposable gloves
Safety goggles
Laboratory apron
Paper plates

Reagents
Ferric ammonium citrate
Potassium ferricyanide
Glacial acetic acid
3% hydrogen peroxide

Safety

1. Wear safety goggles and a lab apron at all times in the laboratory.
2. Acetic acid fumes are irritating. Be sure to keep the acetic acid under an exhaust fan while you are using it. Gloves may be worn to protect your hands when pouring the acetic acid.
3. No food or drink is allowed in the laboratory at any time.

Procedure

Day 1 - Making the Cyanotype Paper

Note: As much as possible, this experiment should be carried out in subdued light.

1. Measure 2.5 g of ferric ammonium citrate and 3.0 g of potassium ferricyanide into a small beaker.

2. Add 25-30 mL of tap water to the beaker. Stir or swirl to dissolve.

3. Add 1 mL (20 drops from a plastic pipet) of concentrated acetic acid to the beaker. Record the color of the solution.
   Note: Be careful to keep the acid under an exhaust hood during this process.

4. Pour the solution into a pie pan. Soak 2-3 pieces of filter paper in the solution.

5. Place the wet pieces of filter paper on a paper plate, separated by pieces of paper toweling. Place them in a dark place, such as a laboratory drawer, to dry overnight.

Day 2 - Cyanotype Photography

1. Select the objects for your " pictures". Any strange or interesting shape will work.

2. Take one or two pieces of the light-sensitive paper (from the dark storage place) outside along with the objects. Place the paper on a piece of cardboard or a paper plate, and place the objects on the paper. Be sure the paper is in an unobstructed, flat, sunny spot. Record the color of your light-sensitive paper.
   Note: Be careful to watch for shadows as they will be recorded on your paper!

3. Leave the objects on the paper in the sun for 15-20 minutes. The longer the paper is exposed, the better your print will be. Look for a change in the appearance of the paper from a green color to a dusty, dark blue. Record the color of your paper just before you bring it inside.

4. Take the paper indoors and immediately remove the objects. Carefully run water from a faucet over the paper until the water runs clear. Record the color of your paper.

5. Immerse your washed print in 3% hydrogen peroxide (in a pie pan) for 10 seconds and then rinse once again with water. This process will darken the image. Record the color of your paper.

6. Leave the sunprinted paper on a paper towel to dry.

Cleaning Up

1. Place your solution from the pie pan into the waste container provided by your teacher. Rinse the pie pan thoroughly with water and dry it.
2. Clean all glassware and return it to its proper location.
3. Place used gloves, paper plates and filter paper in the waste basket.
4. Wash your hands before leaving the laboratory.

Analysis and Conclusions

Complete the **Analysis and Conclusions** section for this experiment either on your Report Sheet or in your lab report as directed by your teacher.

1. Attach one of your prints to your lab report.
2. Record the day, place and time that the print was exposed.
3. Record the length of the time the print was exposed.
4. Record the weather conditions the day the paper was exposed (bright and sunny, slightly overcast, etc.).
5. In general, how can sunlight start a chemical reaction?
6. In the reaction in this experiment Fe^{3+} ions are reduced to Fe^{2+} ions. How does absorbed sunlight promote this process?
7. The Fe^{2+} ions react with the ferricyanide ions to form an insoluble substance called Prussian blue. Why does the hidden part of the paper return to white after the photographic process is complete?
8. Most photographic work using light-sensitive paper must be done in a dark room. Would carrying out this experiment in a brightly-lit room be a problem?
9. How did the quality of your photograph compare to those of others?

Something Extra

1. Design an experiment that uses sunprint paper to test the effectiveness of sunscreen products. After approval from your teacher, try it and see what happens.
2. Design an experiment that uses sunprint paper to test the effect of various types of incident light. Will the experiment work with fluorescent light? incandescent light? a heat lamp?
3. Using chemical resources find the formula of the Prussian blue molecule.

Experiment 47

Flame Tests

Problem
Can we identify an unknown mixture by using a flame test?

Introduction
Flame tests provide a way to qualitatively test for the presence of specific metallic ions. The heat of the flame excites the electrons in the metal ion, and this energy is released as the electrons "fall back" to their ground states. The color we see is a combination of the visible wavelengths of light emitted by the ion.

In this lab you will perform flame tests on seven different metal ions. You will use your observations to identify two unknown solutions.

Prelaboratory Assignment
✓ Read the entire experiment before you begin.
✓ Answer the Prelaboratory Questions.
   1. Why must the nichrome wire be cleaned thoroughly after each flame test?
   2. Why do we see colors in the flame tests, and why are there different colors for different metal ions?

Materials
Safety goggles   Bunsen burner
Lab apron   Wash bottle with deionized water
Nichrome wire loop   6.0 M HCl
Well plate with solutions   Seven watch glasses (or petri dishes)

Solutions of the following salts:
   barium nitrate
   calcium nitrate   potassium nitrate
   copper(II) nitrate   sodium chloride
   lithium nitrate   strontium nitrate

Safety
1. Safety goggles and a lab apron must be worn at all times in the laboratory.
2. Many of these salts are toxic. If you come in contact with any solution, wash the contacted area thoroughly.
3. The 6.0 M HCl is corrosive. Handle it with extreme care.

Procedure

1. Clean the nichrome wire. First, rinse with deionized water. Next, dip the loop into the 6.0 *M* HCl solution. Place the loop into the flame of the Bunsen burner for about a minute. Pay attention to the color of the clean nichrome wire in the flame.

2. Place a small amount of each solution in separate watch glasses.

3. Perform a flame test on each solution by first heating the loop of the nichrome wire in the Bunsen burner. Hold the watch glass with the solution to be tested next to the intake of the Bunsen burner and place the hot loop into the solution. Make careful observations of the flame of the Bunsen burner and record your observations.

4. Clean the wire as described in step 1, and test each of the remaining six solutions separately.

5. Obtain two unknown solutions from your teacher and perform flame tests on each (cleaning the wire between unknowns). Record all observations.

Cleaning Up

1. Clean up all materials (make sure the nichrome wire is clean).
2. Dispose of all chemicals as instructed by your teacher.
3. Wash your hands thoroughly before leaving the laboratory.

Analysis and Conclusions

Complete the **Analysis and Conclusions** section for this experiment either on your Report Sheet or in your lab report as directed by your teacher.

1. How does the flame test provide support for quantized energy levels? Explain your answer.

2. List the metal ions present in your two unknown solutions and provide reasons.

Something Extra

Does the anion in a salt affect the color observed in a flame test? Design an experiment to answer this question, discuss it with your teacher, and try it.

Experiment 48

Electron Probability

Problem
How well can we model an orbital with a dartboard and a dart?

Introduction
According to modern atomic theory, we cannot be sure of the exact location of electrons in an atom. We predict that electrons will be relatively close to the nucleus (because the electrons are negatively charged and the nucleus is positively charged). However, we discuss the "location" of an electron in terms or probability instead of an exact position.

In this activity you will construct and analyze a probability map using a dart and dartboard.

Prelaboratory Assignment
✓ Read the entire experiment before you begin.

Materials
Safety goggles
Darts
Graph paper
Target
Cardboard

Safety
1. Safety goggles must be worn in the laboratory at all times.
2. Be careful to drop the darts toward the target on the floor. Nobody should ever be in the path of a dart.

Procedure
1. Tape the target to the center of the cardboard, place it on the floor, and tape this to the floor.

2. Drop the dart from shoulder height trying to hit the center of the target. Your partner should retrieve the dart and mark the position of the hit with a small x (don't count drops that fall outside the largest circle).

3. Repeat this procedure 99 times for a total of 100 drops.

4. Count the number of hits in each ring and record this number.

Cleaning Up
1. Return all materials to their proper locations.

Data/Observations

Complete the **Data/Observations** section for this experiment either on your Report Sheet or in your lab report as directed by your teacher. Do not write in the table below.

Ring Number	Average distance from target center (cm)	Area of ring (cm^2)	Number of hits in the ring	Number of hits per unit area $(hits/cm^2)$
1	0.5	3.1		
2	1.5	9.4		
3	2.5	16		
4	3.5	22		
5	4.5	28		
6	5.5	35		
7	6.5	41		
8	7.5	47		
9	8.5	53		
10	9.5	60		

Analysis and Conclusions

Complete the **Analysis and Conclusions** section for this experiment either on your Report Sheet or in your lab report as directed by your teacher.

1. Which is the ring with the highest probability of finding a hit?
2. Which is the ring with lowest probability of finding a hit?
3. Construct a graph of the number of hits vs. average distance from the center.
4. Construct a graph of hits per unit area vs. average distance from the center.
5. What does the graph of the number of hits vs. average distance represent? Account for its shape.
6. What does the graph of the hits per unit area vs. average distance represent? Account for its shape.
7. Is the maximum of each graph the same? Explain.
8. How well can we model an orbital with a dartboard and a dart? Specifically, focus on the following:
   a. Compare your target with Figure 11.15 in your text. How is it similar? How is it different?
   b. Why do we predict an electron should be near the nucleus? Why do we expect the dart to land in the center of the target?

Something Extra

How would changing the number of drops affect your results? Do the experiment with 10 drops and with 500 drops and compare your results.

Experiment 49

Dyes and Dyeing
A Practical Application of Bonding Principles

Problem
How does the fabric industry deal with bonding and intermolecular forces involved in fabric dyeing?

Introduction
The purpose of this activity is to introduce you to one example of the importance of molecular structure and polarity in the commercial world. Cloth is made from fibers, with each fiber consisting of millions of long-chain molecules. Some natural fibers, such as silk and wool, are essentially protein molecules. Since proteins are made from amino acids, which have many polar and/or ionic sites on them, the fibers we make from them should have a strong affinity for polar and ionic substances. Because most dye molecules are either very polar or ionic, we would expect them to bond quite readily to a fabric, such as wool. At the other extreme, nylon has very few polar sites so it is very difficult to dye. In between these extremes are fabrics such as dacron and orlon, each with a few polar sites scattered along the chain. As you might predict, these show intermediate attraction for the various dyes. We will use the intensity of color as our measure of the attraction of fabric for dye – the darker the color, the stronger the attraction.

In order to dye the low-polarity fabrics, it is necessary to alter their molecular structure to improve their receptiveness to ionic and polar dye molecules. This is done through a process known as **mordanting** (the term comes from a Latin word, which means 'to bite;' thus, we are helping the dye to "bite into" the fabric). Mordanting involves affixing metal ions or some other ionic species (such as tannic acid) to the polar sites of the fabric. These ions then serve to bind the dye to the fabric, much as glue holds a rubber gasket to a metal fitting. As part of the study, you will investigate the extent to which mordanting improves the dye-holding capability of the six different fibers we will be testing.

The fabric samples that you will use are actually several different types of fibers arranged in a pattern. These strips are used in the garment industry for testing and identification of the various fabrics. From the preceding discussion and your experimental results, you should be able to decide which end is the wool, and you will be able to compare directly the differences in dye characteristics of the different fabrics. Solutions of the dyes will be located on hot plates around the laboratory. The fabric samples are placed in the baths and left for several minutes.

You and your partner will divide the duties between you; in steps 1 and 4, one of you will use Malachite Green, while the other uses Methyl Orange. In similar fashion, one person does step 2 while the other does step 3. Each of you is to keep your own record of observations, which you will share with each other in completing the report.

Prelaboratory Assignment

✓ Read the **Introduction** and **Procedure** before you begin.
✓ Answer the Prelaboratory Questions.

1. Cotton is pure cellulose, a type of carbohydrate. As such, it is essentially a long chain of carbon atoms with —C—O—H groups arranged along the chain.
   a. Would you expect the atoms in the —C—O—H group to be linear or bent? Explain.
   b. Make a sketch showing how hydrogen bonding could occur between the —C—O—H groups on neighboring molecules.

2. Wool is a protein, as is your hair. Proteins contain many — NH groups on a long chain of carbon atoms. What sort of attractive forces occur between proteins and dye molecules? Explain.

3. The **Introduction** and **Procedure** mention that some dyes are polar molecules, while others are ionic. Which would you expect to form the stronger bonds?

Materials

Apparatus
Tongs or forceps
Fabric test strips
Beaker (250-mL)
Safety goggles
Lab apron

Reagents
Dye baths:
  Methyl orange
  Congo red
  Malachite green
  Indigo
Mordant baths:
  $CuSO_4(aq)$
  $FeSO_4(aq)$
  Dilute $HCl(aq)$

Safety

1. Dyes will stain skin just as well as they will fabric.
2. Some of the solutions contain skin irritants, such as acids or bases. Avoid contact.
3. Safety goggles and a lab apron must be worn in the lab at all times.

Procedure

Part 1   Untreated cloth

1. **Direct Dying**

In direct dying the ionic sites of the dye molecule attach themselves to ionic sites (carboxylic acid and amine groups) of the fiber. A piece of test cloth is immersed in the dye bath and maintained at near-boiling for a period of 5-10 minutes. The strip is then removed from the bath and as much as possible of the dye solution is allowed to drain back into the bath. Excess dye is rinsed away and the fabric is allowed to dry. Once the fabric has dried, you can test the color for fastness against washing; cut the strip in half lengthwise and wash one of the half-strips. Dyes to be tested by direct dying are **Methyl Orange** and **Malachite Green** (the same dyes will be tested on mordanted cloth in Part 2).

2. Substantive Dyes

Fabrics such as cotton and rayons have polar sites but no ionic sites, so they do not bind to the colors as well as cotton or silk. One way around this difficulty is to use **substantive dyes**. These are large colloidal molecules are presumed to be bound to the hydroxyl groups of the cellulose structure (cotton is pure cellulose, while rayon is cellulose acetate) by hydrogen bonds. The substantive dye we will use is **Congo Red**.

Cut a test strip in half lengthwise. Immerse both halves in the Congo Red bath for about 10 minutes, then drain and remove the fabric strips to a warm water bath in your 250-mL beaker. Wash in the warm water for as long as the dye is removed. Rinse one of the half-strips in very dilute hydrochloric acid and observe the result. Rinse the acid-treated strip, then wash both with soap.

3. Indigo: a Vat Dye

Indigo, the dye used to dye denim, is not very soluble in water, so it doesn't rinse out easily. On the other hand, indigo is held to the fabric only by relatively weak intermolecular forces, primarily hydrogen bonds. This means that it will rub off fairly easily, as you probably have observed. You may be surprised to see that the indigo bath is not dark blue, the color we normally associate with the dye. In order to make the indigo dissolve in the water, it was reduced in an oxidation-reduction reaction to produce a form of indigo that is water soluble.

Immerse a piece of test cloth in the Indigo bath and boil gently for about 10 minutes. Thoroughly rinse the strip in water then let it dry. After school, take the strip home and cut it in half lengthwise. Wash one half in a mild soap solution, and the other in the same soap solution but with a little bleach added. (For a bathroom sink, about a capful of liquid bleach is ample.)

Part 2   Mordanted Cloth (Both partners should use the same mordanting bath; copper or iron.)

4. Two mordanting baths have been prepared for you: copper(II) sulfate, a source of Cu^{2+} ions; and iron(II) sulfate, a source of Fe^{2+}. Soak a test strip in one of the near-boiling mordanting baths for at least 20 minutes (30 minutes is better), then wring it out over the mordanting bath. Dye the strip using either Malachite Green or Methyl Orange, as in Part 1. Notice the difference between the mordanted cloth and the untreated fabrics. Compare your results with someone who used the other mordanting bath, but the same dye. [In the interest of time, your teacher may provide you with strips of test fabric that have been mordanted with one of the above solutions.]

Cleaning Up

1. Dye baths can be diluted with water and flushed down the sink; your teacher may do this for you since the baths are hot and stain easily.
2. The metal ion solutions should be saved for future use or disposed of by your teacher.
3. The test strips (except for those used in Part 3) should be left to dry overnight on labeled paper towels. Your teacher may ask you to turn them in with your lab report; they are then yours to keep.
4. Wash your hands before leaving the laboratory.

Analysis and Conclusions

Complete the **Analysis and Conclusions** section for this experiment either on your Report Sheet or in your lab report as directed by your teacher.

1. Consult the fabric list supplied by your teacher for the multifiber strips you used. Overall, which type of fabric seems the most effective at binding to the greatest number of dyes? Which is least effective? Account for the differences.

2. Answer either **a.** or **b.**, depending on which step you carried out.
   a. If you did step 2, describe and account for the effect of the acid rinse on your test strip. How might you reverse that change?
   b. If you did step 3, describe and account for the difference between the portions washed with soap and water only and the one washed with soap, water and bleach.

Something Extra

1. In a reference book or on the internet, locate the structure of one of the dyes that you used. Copy the structure for your report and indicate several of the bonding sites that would help the dye molecule adhere to fabrics.

2. Use the internet or a reference book to locate the structural formulas of cellulose, nylon and any sort of protein structure (representing wool and silk). These molecules are large, often containing several thousand atoms. But, because they are polymers, they have variable sizes. Some individual fibers may be twice as long as others. This means that you will not be able to find an exact formula. Show the unit structure for each molecule you found. Identify some of the polar and ionic sites on each molecule where bonding can occur. Use the nylon structure to explain why it is difficult to dye.

Experiment 50

Models of Molecules

Problem
Can we use the molecular formula to predict the shapes of small molecules?

Introduction
The shape of a molecule greatly affects its properties. In this activity you will use clay and toothpicks to determine the shapes of the small molecules represented by the formulas below. You will use toothpicks to represent both bonding pairs and lone pairs of electrons.

HCN H_2CO CCl_2H_2
O_3 H_2S PH_3

Prelaboratory Assignment
✓ Read the entire experiment before you begin.
✓ Answer the Prelaboratory Question.
1. Draw Lewis structures for each of the molecules given in the Introduction. **Note:** In all of the molecules containing carbon, the carbon atom is the central atom.

Materials
Ruler
Protractor
Toothpicks
Modeling clay

Procedure
1. Use toothpicks and modeling clay to make models for each of the molecules given in the Prelaboratory Assignment. Make sure the toothpicks stick out an equal distance from the center clay ball (make sure this distance is the same for all molecules; in this way we can compare measurements for different molecules). Use toothpicks for both bonding pairs and lone pairs.

2. Use the ruler to make sure the ends of the toothpicks are as far apart from each other as possible. Measure and record this distance.

3. Measure and record the bond angles between the toothpicks.

4. Record the electron pair geometry and shape of each of the molecules.

Cleaning Up
1. Take apart your models. Return all materials as instructed by your teacher.
2. Wash your hands thoroughly before leaving the laboratory.

Analysis and Conclusions

Complete the **Analysis and Conclusions** section for this experiment either on your Report Sheet or in your lab report as directed by your teacher.

1. Which shape has the longest distance between the ends of the toothpicks? Which shape has the shortest distance between the ends of the toothpicks?

2. Group together molecules that have the same number of atoms coming off the central atom. Do these molecules have the same geometry? The same shape?

3. Group together models that have the same number of toothpicks coming off the central atom. Do these molecules have the same geometry? The same shape?

4. What do the toothpicks represent?

5. Why do we position the ends of the toothpicks as far apart as possible?

6. Can we use the molecular formula to predict the shape of a small molecule? Explain.

Something Extra

In this activity the greatest number of toothpicks coming off the central atom was four. Make models with five and six toothpicks coming off the central atom. Show these to your teacher.

Experiment 51

Gas Laws and Drinking Straws

Problem
Why can we drink through a straw?

Introduction
You have probably used drinking straws many times, but have you wondered how they work? In this lab we will investigate the science behind the drinking straw.

Prelaboratory Assignment
✓ Read the entire experiment before you begin.
✓ Answer the Prelaboratory Question.

1. Read Section 13.1 in your text and explain how a barometer works in your own words.

Materials
Preform   Modeling clay
Cup   Straws
Transparent tape

Procedure
Part 1
1. Fill a preform with water. Put a drinking straw into the preform and use clay to seal the opening.

2. Try to drink the water. Record your observations. Are you able to drink the water?

Part 2
3. Half fill a cup with water. Place the ends of two drinking straws in your mouth.

4. Submerge the other end of one straw into the water, and leave the other straw out of the water (on the side of the cup).

5. Try to drink the water. Record your observations. Are you able to drink the water?

Part 3
6. Determine the maximum number of straws you can connect together end to end and still drink water. Record this number (Note: you may have to tape the straws together).

7. Measure and record the height.

Part 4
8. Place a drinking straw vertically into a cup that is half-filled with water. Place your finger over the opening of the straw and take the straw out of the water. What happens? Make and record careful observations.

Cleaning Up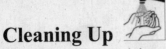

1. Return all materials to their proper locations.
2. Wash your hands thoroughly before leaving the laboratory.

Analysis and Conclusions

Complete the **Analysis and Conclusions** section for this experiment either on your Report Sheet or in your lab report as directed by your teacher.

1. Explain your results to Part 1.

2. Explain your results to Part 2.

3. Compare your number of straws and heights from Part 3 with those of other groups.

4. Why is there a limit to the number of straw through which you can drink?

5. What is the maximum theoretical height through which you can drink? Why can't you drink from this height?

6. Explain your results to Part 4.

7. Why can we drink through a straw?

Something Extra

Does the diameter of a drinking straw affect the results for Part 4? Carry out an experiment to answer this question.

Experiment 52

Determining Absolute Zero

Problem
At what temperature should the volume of an ideal gas reach zero?

Introduction
The volume of a sample of gas kept under constant pressure varies directly with its absolute temperature. In other words, the warmer the gas molecules are, the greater the volume they require at a given pressure. Air confined in a capillary tube follows this law reasonably well. Remember that temperature measures the average kinetic energy of molecules. The only conditions under which the kinetic energy of gas molecules would be expected to be zero is at the temperature called absolute zero. If the gas molecules have zero kinetic energy, they should also have zero volume.

Although it is impossible to actually reach absolute zero, we can determine its value on the Celsius temperature scale. During this investigation you will measure the volume of air in a capillary tube at various temperatures. These volume data will be plotted against the corresponding temperatures. A straight line should result, which you can extrapolate to find the value of absolute zero.

Prelaboratory Assignment
✓ Read the **Introduction** and **Procedure** before you begin.
✓ Answer the Prelaboratory Questions.

1. Under what conditions of temperature and pressure is a gas most ideal?
2. If 740 mL of a gas are collected at 120°C, what volume will the gas occupy if the temperature is decreased to 10°C with the pressure remaining constant?
3. A sample of a gas is collected and measured to be 100.0 mL at standard temperature and pressure. At what Celsius temperature will its volume be twice as great? Half as great? Assume constant pressure.

Materials
Apparatus
Capillary tube
100-mL or 250-mL graduated cylinder
Small rubber band
Thermometer
Hot plate or heating unit
Hot pad
Metric ruler
Safety goggles
Lab apron

Reagents
Ice
Water

Safety

1. Wear safety goggles and a lab apron at all times in the laboratory.
2. No food or drink is allowed in the laboratory at any time.
3. Be careful to hold the capillary tube upright at all times. If the plug in the tube should separate, return the tube immediately to your teacher.
4. Be careful when using the heating apparatus. Be sure to use a hot pad to empty the warm water bath.
5. If you are using a mercury thermometer, be very careful. Mercury vapor is very poisonous. If the thermometer breaks, notify your teacher right away.

Procedure

1. Obtain a capillary tube and use a small rubber band to attach it to a metric ruler so that the bottom of the tube is aligned with the first mark on the ruler and the rubber band is located near the plug of oil. **Note:** Keep the tube upright throughout the experiment.

2. Determine the temperature of the air in the room to the nearest $0.2^{\circ}C$ and record it.

3. Measure the height of the air column in the tube from the sealed end to the base of the plug. Record the height.

4. Fill a beaker 3/4 full with water. Heat the water to boiling. Carefully remove the beaker from the heat. Measure the temperature of the water to the nearest $0.2^{\circ}C$ and record.

5. Immerse the capillary tube in the beaker of hot water ($100^{\circ}C$), taking care not to immerse the rubber band.

6. When the droplet of oil has risen to its highest point, measure and record the height of the column of air in the capillary tube.

7. Repeat steps 4-6 with water baths of approximately 80, 60, 40 and $0^{\circ}C$ (a beaker containing ice and water).

Cleaning Up

1. Empty the water bath and return glassware to its proper location.
2. Return the capillary tube to your teacher, or store it as you are directed.
3. Wash your hands thoroughly before leaving the laboratory.

Analysis and Conclusions

Complete the **Analysis and Conclusions** section for this experiment either on your Report Sheet or in your lab report as directed by your teacher.

1. Obtain a large piece of graph paper. You might cut a standard 8 1/2"x11" piece of graph paper lengthwise and tape the ends together, creating a 4 1/4" by 22" piece of paper.

2. Draw a horizontal axis (along the long side) near the bottom of the paper. The horizontal axis must allow for temperatures ranging from -300°C on the left to 100°C on the right. Locate the vertical axis at 0°C.

3. Plot the temperatures on the horizontal axis and the height of the air column on the vertical axis.

4. Draw the best straight line that fits the points. Extend your line (extrapolate) to the horizontal axis (where the height of air in the tube would be zero).

5. Identify the temperature where the line crosses the horizontal axis. What is your estimated value for absolute zero?

6. By how many degrees does your value differ from the accepted value? Is your answer too high or too low?

7. Should the slope of your line decrease or increase to make your results more accurate?

Something Extra

1. The diameter of your capillary tube actually increases as the tube is heated. Does this affect your graph? Explain.

2. In this experiment you are plotting height versus temperature rather than volume versus temperature. Explain why treating your data this way should still result in the correct value for absolute zero.

3. Air is not an ideal gas. Would air actually have zero volume at absolute zero? Explain.

Experiment 53

The *P-n* Relationship for Gases

Problem

How is the pressure exerted by a gas related to the number of molecules present?

Introduction

In Chapter 13, you are introduced to the four variables that determine the properties of a gaseous system: pressure (P), volume (V), absolute temperature (T), and the number of moles of gas molecules present (n). This experiment will provide you with an opportunity to explore the properties of gases and develop a relationship between pressure and moles of a gas for yourself. You will measure the pressure exerted in a fixed volume by varying number of moles of gas, then use your results to arrive at a general statement concerning how the number of gas molecules in a system affects the pressure the gas exerts.

Baking soda, sodium bicarbonate ($NaHCO_3$), reacts with the acetic acid in vinegar to produce carbon dioxide gas and an aqueous solution of sodium acetate:

$$NaHCO_3(s) + CH_3COOH(aq) \rightarrow Na^+CH_3COO^-(aq) + H_2O(l) + CO_2(g)$$

In this experiment a known mass of sodium bicarbonate is allowed to react with an excess of acetic acid (vinegar) to generate carbon dioxide. The pressure exerted by the carbon dioxide will be measured using a pressure sensor. Knowledge of the mass, and therefore, moles of sodium bicarbonate allows the calculation of moles of $CO_2(g)$ produced. This, in turn, allows calculation of the molar volume of carbon dioxide, i.e. the volume in liters occupied by one mole of CO_2 at standard temperature and pressure. The process will be carried out a total of four times, with increasing masses of baking soda.

Prelaboratory Assignment

✓ Read the **Introduction** and **Procedure** before you begin.
✓ Answer the Prelaboratory Questions.

1. Calculate the mass of baking soda that would generate enough carbon dioxide to exert 1.0-atm pressure in a 125-mL flask at 20°C. This is the maximum amount of baking soda that will be used in this experiment. Show your calculations.

2. Look up the solubility of carbon dioxide in water at 25°C and one atmosphere pressure in *The CRC Handbook of Chemistry and Physics*. Assuming that such a solution would have a density of 1.00 g/mL, calculate the molar solubility (mol CO_2/L of solution) of carbon dioxide at 25°C.

3. **a.** Calculate the number of moles of carbon dioxide gas that would dissolve in 5.0 mL of water at 25°C. Assume the same molar solubility in vinegar as you calculated for water in Question 2. Show your work.

   b. What pressure would be exerted in 125-mL at 25°C by the number of moles of CO_2 that you calculated in **3a**? This is the amount by which the pressure changes that you observe in the experiment will be reduced because some of the $CO_2(g)$ produced dissolves in the vinegar solution. Show your work.

Materials

Apparatus
Milligram balance
Graphing Calculator (TI83+ or similar)
CBL2, LabPro or similar interface
Vernier Pressure Sensor with stopper, connector tube and syringe
    or other interface-specific pressure sensor
    (must be able to withstand about 2.0 atm)
125-mL Erlenmeyer flask
Graduated cylinder, 100-mL or larger(optional)
Safety goggles
Lab apron

Reagents
Sodium bicarbonate
White vinegar

Safety

1. Safety goggles and a lab apron must be worn at all times in the laboratory.

Procedure

LabPro and CBL2 both use the DataMate application. If you have not yet loaded the appropriate program into your calculator, you must do so before you begin. You can either obtain the program from another calculator of the same model, or you can transfer DataMate directly from either the Vernier or Texas Instruments interface.

Setting Up the System

1. Connect the calculator to the interface using the link cable, then attach the pressure sensor to Channel 1 of the interface. Turn on the calculator and CBL. After a pause, you should see CHECKING SENSORS on the calculator screen. Both CBL2 and LabPro will recognize the Vernier Pressure Sensor. If you wish to change the pressure units, use the SETUP menu. You can also use the SETUP menu to set the MODE to EVENTS WITH ENTRY.

   The apparatus you will use consists of a 125-mL Erlenmeyer flask, the Vernier pressure sensor and a syringe. A known mass of sodium bicarbonate is placed directly in the bottom of the flask. The vinegar is drawn into the syringe that comes with the pressure sensor. The syringe is screwed into the fitting above the stopcock. When the vinegar is injected into the flask, it begins to react with the baking soda. As $CO_2(g)$ is formed, it causes the pressure in the flask to increase. By varying the amount of sodium bicarbonate, you can establish a relationship between moles of carbon dioxide generated and the pressure exerted, as recorded by the pressure sensor.

2. Use the piece of clear plastic tubing that comes with the pressure sensor to connect the shorter fitting on the stopper to the pressure sensor. Clamp the flask to a ring for stability, or have one partner hold it while the other makes the remaining connections. Be certain that the stopcock is closed.

3. On a piece of weighing paper, measure about 0.05 g (±1 mg) of sodium bicarbonate, record the mass and transfer it to the flask.

    Note: If the flask has been previously used, for this or another experiment, rinse it well with distilled water and shake out excess water before adding the solid. It does not need to be dried.

4. Fill the syringe with vinegar to the 5 mL mark. Fit the syringe snugly into the stopcock on the stopper. Do not open the stopcock on the valve yet.

5. Press 2-START to have the CBL monitor the pressure in the flask. When the pressure reading is steady, press ENTER; when you are asked to enter a VALUE, enter 1 and press ENTER once again.

    Now open the stopcock, but do not add the vinegar. Press ENTER again; the VALUE for this reading is "2." The chances are, the first two readings will be the same.

6. Now firmly, but smoothly inject the vinegar into the flask. While you are holding the plunger all the way down, close the stopcock. You will need to maintain a firm hand pressure on the stopper.

7. When all bubbling stops, the reaction is complete. You will also see that the pressure levels off and remains constant. Use ENTER to collect your third and last data point. This time you want to STOP (STO). The calculator will display the graph on the screen; use the left and right arrows to determine the initial and final pressures. The difference between the final and initial pressures is the pressure of the CO_2 produced in the reaction.

8. Remove the stopper from the flask. Rinse the flask, first with tap water, then with distilled water, and dry the inside as well as you can. Repeat the experiment two or three more times, increasing the mass of baking soda each time. Each mass should be 0.05-0.10 g greater than the preceding mass.

    Note: Do not exceed the mass of $NaHCO_3$ that you calculated in Prelaboratory Question 1. Be aware that the more gas you generate, the more likely the system is to leak.

9. **Optional.** The 125-mL volume of the flask is approximate. You might want to determine the actual volume, using a graduated cylinder. If you do, be sure to reduce the volume you measure by the 5.0 mL occupied by the vinegar. Record this volume in your observations.

Cleaning Up

1. At the conclusion of each trial, the contents of the flask may be rinsed down the drain with water.
2. Return the equipment to the locations specified by your teacher.
3. Wash your hands before leaving the laboratory.

Analysis and Conclusions

Complete the **Analysis and Conclusions** section for this experiment either on your Report Sheet or in your lab report as directed by your teacher.

If you are not using a Report Sheet for this experiment make a Summary Table with the following headings:

Trial	Mass NaHCO$_3$(s)	Initial Pressure	Final Pressure	Pressure CO$_2$

1. For each trial you conducted, use the mass of sodium bicarbonate to calculate the moles of CO$_2$ generated. Show your work for Trial 1 and list the results for all trials.

2. On graph paper, make a plot of pressure of CO$_2$ *versus* moles of CO$_2$. How closely does your line fit the ideal gas assumption that when $n = 0$, $P = 0$? Discuss. Attach the plot to your report.

3. Convert your results from **1** to values for the ratio of pressure to moles of CO$_2$ for each trial, in L/mol. Calculate an average value, including average deviation. Show one sample calculation.

Something Extra

1. What you have done so far assumes that all of the carbon dioxide was present in the space above the mixture in the flask. In Prelaboratory Question **3a**, you calculated the amount of CO$_2$ that would dissolve in the liquid in your flask. In **3b**, you determined the amount by which the pressure readings would be in error.
   a. Adjust each of your values for the observed pressure of CO$_2$ by subtracting this pressure.
   b. Re-plot pressure *vs.* moles, as described in Question 2. Discuss the effect(s) this correction has on your results for Questions 2 and 3 of Analysis and Conclusions.

2. In reality, the solubility varies as the pressure of the CO$_2$ increases. This variation in solubility with pressure is known as Henry's Law. Consult an advanced text to see how you can determine the actual solubility for each of your experiments, then recalculate the corrected pressures that you found in the preceding question. Discuss the significance of your results.

Experiment 54

Molar Volume and the Universal Gas Constant

Problem

How can the value of the universal gas constant be verified experimentally?

Introduction

There are a number of gas laws described in Chapter 13. Some, like the laws of Avogadro, Boyle and Charles, concern how one of the four variables – pressure, volume, absolute temperature, and number of moles – varies when another is changed, with two of the four being held constant. Two other relationships are of a more general nature. One, the Combined Gas Law, represents a combining of the simpler laws. Given that volume is inversely proportional to pressure (Boyle's Law), and directly proportional to temperature (Charles' Law), you can write an expression that incorporates both sets of observations:

$$P_1V_1/T_1 = P_2V_2/T_2$$

Going one step further and introducing the law of Avogadro, which says that the volume occupied by a gas is directly proportional to the number of moles of gas in the system, gives the expression

$$P_1V_1/ n_1T_1 = P_2V_2/n_2T_2$$

In which P_1, V_1, n_1 and T_1 refer to an initial set of conditions and P_2, V_2, n_2 and T_2 refer to the final conditions, after the change has occurred. This "combined" gas law allows you to do calculations involving changes in any or all of the four variables.

What the equation tells us is that the value of the expression, PV/nT, does not change; if that is the case, then we can determine a value for PV/nT that should be valid under any set of conditions as long as the sample remains a gas. In other words we can state that

$$PV/nT = \text{constant} = R$$

This relationship ($PV/nT = R$) can be rearranged to the form known as the Ideal Gas Law:

$$PV = nRT$$

The constant, R, is known as the *universal gas constant*. One of your objectives in this experiment is to determine the value of R experimentally. Notice that the units of R must reflect pressure times volume, divided by number of moles times temperature. R is most often expressed in units of L atm/mol·K ("liter-atmospheres per mole-kelvin"). In this experiment you will calculate the value of R for each of three trials, as well as an average result which will then be used to determine your percentage error.

You will first determine the molar volume of a gas. You will then use the molar volume at laboratory conditions to determine what volume one mole of gas would occupy at STP ($0°C$ and one atmosphere pressure). You can make this conversion by using the combined gas law. Once that is done, you will use your experimental values of *P, V, n,* and *T* to calculate an experimental value for *R*. As with the molar volume calculation, you will determine individual values for each trial, along with an average value, which you will compare with the accepted value of *R*: 0.0821 L atm/mol·K.

Prelaboratory Assignment

✓ Read the **Introduction** and **Procedure** before you begin.
✓ Answer the Prelaboratory Questions.

1. Explain how the mass of a piece of magnesium ribbon may be calculated from its length and the mass of a 100.0-cm strip of the same type of ribbon.

2. What two physical properties of hydrogen gas make it possible for you to collect it by displacement of water in your graduated cylinder?

3. Use Dalton's Law of Partial Pressures to explain why the pressure of hydrogen gas in the cylinder will be less than the observed barometric pressure in the laboratory. How will you determine the pressure of the hydrogen you produce?

4. Calculate the mass of baking soda, $NaHCO_3$, needed to neutralize 3.0 mL of 3.0 *M* hydrochloric acid, HCl(*aq*). Show your work; the answer alone is not enough. (**Hint:** the amount needed is less than 1 gram.)

5. The accepted value for the universal gas constant is 0.0821 L atm/mol·K. What would it be if the pressure was measured in torr and the volume in milliliters? Show your calculations.

Materials

Apparatus
10-mL graduated cylinder
#00 1-hole rubber stopper
Copper wire, 10 cm length
Thermometer
Safety goggles
Lab apron

Reagents
Magnesium ribbon, 3 pcs, ~0.9 cm
3 *M* HCl (*aq*)
$NaHCO_3(s)$ (baking soda)

Safety

1. Wear safety goggles and a lab apron at all times in the laboratory.
2. Hydrochloric acid is corrosive to skin and clothing. Clean up all spills thoroughly.
   Note: If acid spills on the lab bench, use a bit of baking soda to neutralize it before cleaning up.
   Do not neutralize acid that spills on skin or clothing; flood the affected area with water.
3. If you are using a mercury thermometer, be very careful. Mercury vapor is very poisonous. If the thermometer breaks, notify your teacher right away.

Procedure

1. Fill a 400-mL beaker about three-fourths full of tap water; the water should be at or near room temperature.

2. Obtain a short (0.80-1.00 cm) piece of magnesium ribbon, then measure and record its length to the nearest 0.01 cm. Use a 10-cm piece of copper wire to make a cage for your magnesium, by folding the magnesium over the wire then rolling the wire around the magnesium. Fit the wire cage into a 1-hole #00 rubber stopper. The cage should be about 2-3 cm from the small end of the stopper to hold the magnesium in place. See **Figure 1**.

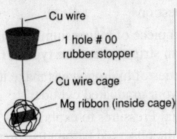

Figure 1

3. Carefully pour about 3 mL of 3 M HCl(aq) into a 10-mL graduated cylinder. Using a wash bottle or beaker, carefully add distilled or deionized water to the graduated cylinder until it is completely full. Try to direct the water down the side of the graduated cylinder to prevent mixing of the water and the acid. Insert the stopper assembly (see **Fig. 1**, above) into the top of the graduated cylinder. Water should escape through the hole in the stopper; if it does not, remove the stopper and carefully add more water, then replace the stopper assembly. This will keep air from being trapped in the graduated cylinder.

4. Place your finger over the hole in the stopper and invert the graduated cylinder, lowering it into the beaker of water. Remove your finger when the stopper is below the level of water in the beaker. The hydrochloric acid is more dense than pure water, so it will slowly sink toward the stopper and the magnesium; observe and record evidence of reaction. (See **Figure 2**.)

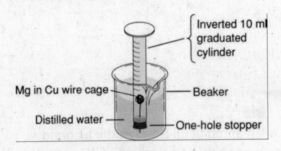

Figure 2

5. When the reaction is complete, allow the system to stand for two or three minutes, tapping the sides of the graduated cylinder to dislodge any gas bubbles that may be clinging to the glass wall. Make sure that there are no little pieces of unreacted magnesium on the wall of the graduated cylinder. If a small piece remains, gently shake the graduated cylinder up and down to wash the metal back into the acid solution, allowing it to finish reacting. Be careful not to lift the graduated cylinder completely out of the water in the beaker.

6. Adjust the position of the graduated cylinder so that the water levels inside and out are even; this ensures that the total pressure on the gases inside the graduated cylinder is the same as the barometric pressure in the room. Record the volume of gas trapped in the graduated cylinder, and the temperature of the water near the mouth of the graduated cylinder. The temperature of the escaping solution may be assumed to be the same as the temperature of the trapped gases. Enter these data in the Data Table.

Cleaning Up

1. Take apart the apparatus. Dispose of the copper wire in the solid waste container or as your teacher directs.
2. Use baking soda to neutralize the acidic solution remaining in the beaker. The neutralized solution can be flushed down the drain safely. As part of the Prelaboratory Assignment, you calculated the mass of baking soda needed. Use a plastic spoon to add approximately that amount of the solid (a little at a time to minimize foaming) to the beaker. Stir well and flush the solution down the drain.
3. Clean all glassware and return it to its proper location.
4. Wash your hands before leaving the laboratory.

Analysis and Conclusions

Complete the **Analysis and Conclusions** section for this experiment either on your Report Sheet or in your lab report as directed by your teacher.

All of the calculations are to be shown for your first trial; you may simply report the results for the other two. Enter the results for all trials in a Summary Table.

1. Calculate the mass of magnesium for each trial, using the mass of 1.00 m of magnesium ribbon supplied by your instructor.

2. Calculate the number of moles of magnesium used. This is the same as the number of moles of hydrogen generated. (Why?)

3. Because you collected hydrogen over water, a small portion of the gas in the cylinder at the end of the reaction is water vapor; we say the hydrogen gas is "wet." The amount of water that evaporates is dependent only on the temperature, so it is a simple matter to determine the *partial pressure* of the water vapor in the graduate. Use the table of vapor pressures, found in Appendix A of your lab manual, to find the pressure due to water in the graduated cylinder. Subtract this value from the barometric pressure to get the pressure exerted by the "dry" hydrogen (hydrogen without the water vapor). Be sure to report the pressure to the correct degree of precision. Enter the results in Summary Table 1.

4. You have measured the volume occupied by a very small fraction of a mole of hydrogen, under a specific set of conditions of pressure and temperature. The volume occupied by one mole of gas is called the **molar volume** of the gas, and it is the same for all gases (behaving ideally) at a particular pressure and temperature. For each trial:

   a. Calculate the volume that 1.00 mole of hydrogen would occupy at your experimental temperature and pressure (called "laboratory conditions"). Record your answers in the Summary Table.

   b. Use the combined gas law to calculate the volume that 1.00 mole of hydrogen would occupy at 1.00 atm and 273 K (Standard Temperature and Pressure, STP).

5. Determine the value of PV/nT for each trial. P is the pressure of dry hydrogen, V is the volume of gas collected, T is the Kelvin temperature, and n is the number of moles of hydrogen generated.

6. Determine the average of your three values for PV/nT. Also determine the deviation for each of your three trials and the average deviation.

7. Your average for PV/nT represents your experimental value for the universal gas constant, R. Calculate the percent error in your determination of the value of R.

Experiment 55

Magic Sand

Problem

How is Magic Sand different from regular sand? What are differences in the interactions of the two Types of sand with other substances?

Introduction

Beach or playground sand is composed of silica that has been broken into small pieces (grains). A model of silica is shown in **Figure 1**.

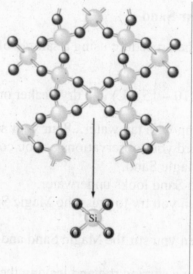

Figure 1

The surface of sand grains contain many oxygen atoms that can hydrogen bond to water molecules.

Magic Sand is also sand, but it has been specially treated to color it, and also to reduce the hydrogen bonding interactions between it and other substances. In this experiment you will have an opportunity to compare sand and Magic Sand to see how they are similar and how they differ.

Prelaboratory Assignment

✓ Read the **Introduction** and **Procedure** before you begin.
✓ Answer the Prelaboratory Questions.

1. What do we mean when we say that water "wets" a surface such as the surface of sand?

2. What is the difference between the way water "wets" the surface of a ten-year old car that has never been waxed, and the same car after it has been freshly waxed?

3. Water and sand can both form hydrogen bonds. How might this explain why wet sand clumps together?

Materials

Magic Sand
4 small beakers or
    clear plastic cups
Dropper
Small paper cup
Safety goggles
Lab apron

Regular sand
Graduated cylinder
Stirring rod
Food coloring
Liquid dishwashing detergent

Safety

1. Wear safety goggles and a lab apron at all times in the laboratory.

Procedure

Part 1   Magic Sand *vs.* Regular Sand

Complete each step in this part of the procedure using a sample of Magic Sand and then repeat the same step using regular sand.

1. Place a sample of Magic Sand (10 – 15 mL) in a dry beaker or cup. Record your observations.

2. Half - fill an empty beaker or cup with tap water. Pour your sample of Magic Sand into the beaker or cup all at once. Record your observations as you complete the following:
   a. Describe the shape of the Magic Sand.
   b. Describe the way the Magic Sand looks underwater.
   c. Describe what happens when you try to press the Magic Sand into different shapes under water.
   d. Describe what happens when you stir the Magic Sand and water with a stirring rod.

3. Carefully decant the water using a stirring rod and leaving the Magic Sand in the beaker or cup.
   a. Describe the shape of the Magic Sand.
   b. Describe the surface of the Magic Sand.

4. Repeat Steps 1 – 3 using regular sand in place of Magic Sand.

Part 2   Properties of Magic Sand

1. Half-fill a cup or beaker with tap water. When ready sprinkle a thin layer of Magic Sand into the beaker or cup containing the tap water. Record your observations.

2. Half-fill a small paper cup with water. Add a drop of food coloring that is different from the color of your Magic Sand sample. Using a dropper add a drop of this colored water to the surface of the Magic Sand from step 1 and observe what happens. Continue to add drops and record the maximum number of drops you can add to the Magic Sand's surface.

3. Half-fill a small cup or beaker with water. Add several drops of dishwashing liquid and stir gently to mix. Pour a small amount of Magic Sand into the cup or beaker. Record your observations.

Cleaning Up

1. Decant the detergent solution from the Magic Sand. Rinse it with water to remove all traces of detergent and return the Magic Sand and regular sand to your teacher.
2. Before you return the Magic Sand from Part 1, rinse off the food coloring.
3. Clean up your station and wash your hands before leaving the laboratory.

Analysis and Conclusions

Complete the **Analysis and Conclusions** section for this experiment either on your Report Sheet or in your lab report as directed by your teacher.

1. Write a paragraph explaining how Magic Sand differs from regular sand in its properties.

2. How did adding dishwashing detergent change the properties of Magic Sand? Explain.

3. What is the major type of intermolecular forces present in water? How does Magic Sand interact with molecules that exhibit this type of intermolecular forces? How does regular sand interact with molecules that exhibit this type of intermolecular forces?

Something Extra

Predict what would happen if Magic Sand were placed in vegetable oil. Explain why this would be different from what happened in this experiment. Try it and see if your prediction is correct.

Experiment 56

Freezing Point – A Physical Property

Problem
What is the freezing point of pure acetic acid? How is the freezing point of a pure substance related to its melting point?

Introduction
Every chemical substance has a set of unique physical and chemical properties. This set of properties determines a fingerprint of the substance that can be used to distinguish the substance from the many hundreds of thousands of other pure substances known to us.

Melting or freezing point and boiling point are properties that are easily determined and very helpful in identifying a substance. These properties are virtually always described in the chemical literature and in computer data bases.

Prelaboratory Assignment
✓ Read the Introduction and **Procedure** before you begin.
✓ Answer the Prelaboratory Questions.

1. What is the difference between a physical property and a chemical property?

2. On a molecular level what is occurring when a substance is cooled to the freezing point?

Materials

Apparatus
Large test tube (18 x 150)
Slotted cork or stopper
    to fit the test tube
Ring stand
10-mL graduated cylinder
Lab apron
Safety goggles
Gloves (optional)
400-mL beaker
Thermometer
Utility clamp

Reagents
Pure acetic acid

Safety

1. Pure acetic acid is corrosive to skin and clothing. You may want to wear gloves when handling this material. Clean up spills with large amounts of water.
2. Wear laboratory safety goggles and a lab apron at all times in the laboratory.
3. If you are using a mercury thermometer; be very careful. Mercury vapor is very poisonous. If the thermometer breaks, notify your teacher right away.

Procedure

If you are not using a Report Sheet for this experiment make a Data Table with headings for Time and Temperature.

1. Attach the test tube clamp to the ring stand so the tube is about 25 cm above the base. Place the test tube in the clamp near the top of the tube and tighten the clamp securely (without breaking the test tube).

2. Place the thermometer in the slotted cork to that the temperature scale shows in the slot. Position the cork so that the thermometer is about 1 cm from the bottom of the test tube when the cork is in place as shown in Figure 1.

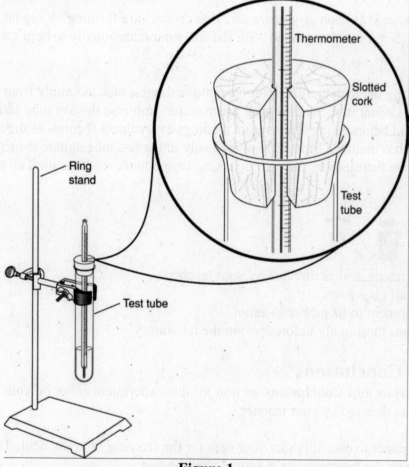

Figure 1
Apparatus for measuring temperature

3. Measure 10.0 mL of pure acetic acid using a graduated cylinder. Pour the acid into the test tube and replace the thermometer.

4. Adjust the temperature of the acetic acid to about 25°C by warming or cooling the test tube in a beaker of water.

5. Empty the beaker and fill it about ¾ full of crushed ice. Add cold water until the ice is almost covered. Place the beaker of ice water below the test tube assembly.

6. Carefully read the thermometer and record the temperature as the 0 minute reading in your data table.

7. Lower the test tube assembly into the ice water. Be sure to lower the test tube until all of the acetic acid is under water. Refasten the clamp to the ring stand.

8. Loosen the cork slightly and agitate the acid slightly using the thermometer. Be careful not to bang the thermometer bulb on the glass and keep the thermometer bulb completely immersed in the acid. Take temperature readings every 30 seconds as the acid cools. Begin with the 0.5 minute reading.

9. Stop stirring the acid as soon as you are sure that crystals are forming. Keep taking readings until a total of 15 minutes has passed. Stir the ice water occasionally to help keep it at a constant temperature.

10. After completing the temperature readings remove the test tube assembly from the ice water. Replace the ice water in the beaker with warm water. Immerse the test tube assembly in the warm water and begin taking temperature readings every thirty seconds as the acetic acid melts. Once the acid has melted a little and moves easily in the test tube agitate the solid-liquid mixture with the thermometer. Continue taking temperature readings until all of the acetic acid is melted.

Cleaning Up

1. Dispose of the acetic acid as directed by your teacher.
2. Clean and dry all glassware.
3. Return all equipment to its proper location.
4. Wash your hands thoroughly before leaving the laboratory.

Analysis and Conclusions

Complete the **Analysis and Conclusions** section for this experiment either on your Report Sheet or in your lab report as directed by your teacher.

1. Use graph paper to carefully plot your data for the freezing of acetic acid. Draw a smooth curve through the freezing data points. Label the line.

2. Use the same graph paper to carefully plot your data for the melting of acetic acid. Draw a smooth curve through the melting data points. Label the line.

3. What is the freezing point of pure acetic acid? How can you tell?

4. What is the melting point of pure acetic acid?

5. How do the freezing point and melting point of pure acetic acid compare?

Something Extra

What happens to the freezing point of a pure substance when another substance is added to it? Add 0.5 g of benzoic acid crystals to your pure acetic acid and repeat the experiment. Was your prediction correct? Give an example of how this process can be used in everyday life.

Experiment 57

Heating and Cooling Behavior of a Pure Substance

Problem
What is the freezing point of *para*-dichlorobenzene crystals? How is the freezing point of a pure substance related to its melting point? What temperature changes are observed when a pure substance warms or cools?

Introduction
Every substance has a set of unique physical and chemical properties. These properties determine a fingerprint of the substance that can be used to distinguish that substance from the many hundreds of thousands of other pure substances known to us.

Melting or freezing point and boiling point are properties that are easily determined and that are very helpful in identifying a substance. These properties are virtually always described in the chemical literature and reference tables.

In this lab you will use the pure substance *para*-dichlorobenzene and investigate its melting and freezing points by simply warming and cooling a sample of the substance. The data you gather will reveal an interesting behavior common to all pure substances.

Prelaboratory Assignment
✓ Read the **Introduction** and **Procedure** before you begin.
✓ Answer the Prelaboratory Questions.

1. What is the chemical structure of *para*-dichlorobenzene?
2. On a molecular level, what is the difference between a substance melting and a substance freezing? Between a substance melting and a substance vaporizing?
3. What happens to the molecules of a substance as it warms? What type of energy is involved?

Materials
Apparatus
Large test tube (18 × 150 mm)
Heating unit
Utility clamp
Stop watch or classroom clock
2 thermometers
Safety goggles
Lab apron
400-mL beaker

Reagents
Para-dichlorobenzene crystals
Acetone

Safety

1. Wear safety goggles and a lab apron at all times in the laboratory.
2. No food or drink is allowed in the laboratory at any time.
3. Be cautious when handling the warm *para*-dichlorobenzene. Do not smell its fumes directly; do not heat excessively.
4. If using a mercury thermometer be very careful. Mercury vapor is poisonous. If you break a mercury thermometer notify your teacher immediately.
5. Remember that when heating a substance in a test tube you should never point the open end of the tube toward anyone.

Procedure

If you are not using a Report Sheet make a Data Table with headings for Time, Temperature (*para*-dichlorobenzene), and Temperature (water).

Part 1 Preparing the *Para*-dichlorobenzene

1. Fill a 400-mL beaker about 3/4 full of cool tap water. Place a thermometer in the beaker and record the temperature. **Note:** Throughout this experiment, you should try to read temperatures to the nearest 0.2°C.

2. Obtain a test tube containing solid *para*-dichlorobenzene from your teacher.

3. While holding it with a utility clamp, heat the tube gently over a medium Bunsen burner flame just until the solid melts. **Note:** Do this carefully and slowly. Move the test tube in and out of the flame at a slight angle. Do not heat excessively to a boil.

4. Once the *para*-dichlorobenzene has melted, remove it from the flame.

5. Decide which partner will stir and make thermometer readings while the other keeps time, records the temperature readings, and makes observations.

Part 2 The Cooling Behavior of *Para*-dichlorobenzene

1. Carefully insert a thermometer into the test tube of melted *para*-dichlorobenzene. Record the temperature.

2. Lower the test tube into the beaker of water. Be sure to lower the test tube until all of the *para*-dichlorobenzene is under water. Fasten the clamp to the ring stand.

3. Wait 30 seconds. During that time you should stir the *para*-dichlorobenzene by moving the thermometer up and down. At the end of 30 seconds, record the temperature of the *para*-dichlorobenzene.

4. Wait another 30 seconds. Stop stirring the *para*-dichlorobenzene as soon as the thermometer begins to stick in the solidifying substance. Record the temperature again.

5. Continue taking temperature readings of the *para*-dichlorobenzene every 30 seconds. Also take temperature readings of the water in the beaker every 60 seconds. Stop timing when the temperatures of the *para*-dichlorobenzene and the water are within 3-5 degrees of each other.

6. After completing this part of the experiment, remove the test tube from the water, and leave it clamped to the ring stand. Leave the thermometer stuck in the test tube.

Part 3   Warming Behavior of *Para*-dichlorobenzene

1. Fill a 400-mL beaker about 3/4 full with hot tap water. Place it on a heating unit and adjust its temperature to about 70°C.

2. Lab partners should exchange duties at this time. Record the temperature of the frozen *para*-dichlorobenzene in the test tube to the nearest 0.2°C.

3. Remove the warm water bath from the heating unit. On the signal of the recorder, lower the test tube into the warm water until the level of the water is above the level of the solid *para*-dichlorobenzene.

4. Clamp the test tube in place. Wait 30 seconds. As before, record the temperature of the *para*-dichlorobenzene at the end of 30 seconds.

5. Record the temperature of the *para*-dichlorobenzene every 30 seconds; record the temperature of the water in the beaker every 60 seconds.

6. As soon as the solid is freely moving in the test tube, move the thermometer gently and continuously up and down.

7. Continue to stir and to record temperatures until the water and the *para*-dichlorobenzene are again within 3-5 degrees of each other. Remove the thermometer from the liquid and then remove the test tube from the water bath.

Cleaning Up

1. Clean the thermometer by dipping it into an acetone solution provided by your teacher.
2. Allow the test tube to cool and the *para*-dichlorobenzene to resolidify. Return the tube containing the *para*-dichlorobenzene to your teacher.
3. Finish cleaning up your station and return all glassware to its proper location.
4. Wash your hands thoroughly before leaving the laboratory.

Analysis and Conclusions

Complete the **Analysis and Conclusions** section for this experiment either on your Report Sheet or in your lab report as directed by your teacher.

1. Use a piece of graph paper to plot all of your data. Plot temperature along the vertical axis and time along the horizontal. Carefully calibrate your axes to use as much of the paper as possible.

2. Use a dot in a small circle for each point of the cooling data, and use a dot in a triangle for each point of the warming data. Use different colors for the water temperature data.

3. Plot all four lines of data on the same sheet of graph paper.

4. Draw a smooth curve through the warming data points. Draw a smooth curve through the cooling data points. Draw a smooth curve for each of the water temperature lines as well.

5. Label each line on your plot.

6. Describe the general appearance of your lines. What general trends do you notice?

7. From your graph, what is the freezing point of *para*-dichlorobenzene? How can you tell?

8. From your graph, what is the melting point of *para*-dichlorobenzene?

9. How do the freezing point and melting point compare?

10. What effect would increasing the amount of solid have on the shape of the melting or cooling graph?

Something Extra

1. Use chemical resources to find the true melting point and freezing point of *para*-dichlorobenzene. How does your experimental result compare?

2. What is the significance of plotting the data for the water in both processes? Why does the *para*-dichlorobenzene data show different temperature behavior than the water lines?

3. What if pure substances are not really pure; that is, what if they are contaminated? What impact does the contamination of a pure substance have on its melting or freezing points?

Experiment 58

Heat of Fusion of Ice

Problem
How much heat is required to melt a gram of ice?

Introduction
Ice and water can coexist at the freezing point. To melt the ice energy must be added. In this lab you will determine the amount of heat required to melt a gram of ice.

Prelaboratory Assignment
✓ Read the entire experiment before you begin.
✓ Answer the Prelaboratory Questions.

1. You could have made volume measurements of the water and used the density of water as 1.0 g /mL to make your calculations. Why would this introduce error?

2. Why must excess ice be added? What problems might occur if this ice were not in excess?

Materials
Apparatus
Safety goggles
Lab apron
Styrofoam cups (2)
Lid for cup (with 2 holes)
Plastic spoon
Stirrer
Thermometer

Reagents
Ice chips (or small cubes) at 0°C
Water

Safety
1. Thermometers are fragile. Be careful in handling them and never use a thermometer as a stirring rod. If you are using a mercury thermometer and it breaks, notify your teacher immediately. Mercury vapors are poisonous.
2. Safety goggles and a lab apron must be worn in the laboratory at all times.

Procedure
1. Place one Styrofoam cup into the other. This is your calorimeter.

2. Place 75.0 g of water at room temperature in the calorimeter. Record the temperature of the water.

3. Using a paper towel, remove the residual water from several ice chips and add them to the water in the calorimeter. Cover the calorimeter.

4. Gently stir the water as the ice melts. Do not stir too vigorously or you will affect the temperature of the water. Add ice if necessary (there should always be ice present).

5. When the temperature of the ice-water mixture is 0°C, use a plastic spoon to remove any excess ice in the calorimeter. Be careful not to take any water.

6. Measure and record the final mass of the water.

Cleaning Up

1. Return all materials to their proper locations. Your teacher will tell you where to place the foam cups.
2. Wash your hands thoroughly before leaving the laboratory.

Analysis and Conclusions

Complete the **Analysis and Conclusions** section for this experiment either in your Report Sheet or in your lab report as directed by your teacher.

1. Determine the heat transferred from the water to the ice.

2. According to your results, how much heat is required to melt a gram of ice?

3. The actual value for the heat of fusion of ice is about 330 J/g. Determine the percent error in your value.

4. Would each of the following scenarios cause you to calculate a heat of fusion higher or lower than the accepted answer? Explain your answer.
   a. The ice you add to the calorimeter has a significant amount of water on it.
   b. The ice is initially colder than 0°C.
   c. A significant amount of water is taken when the excess ice is removed from the calorimeter.

Something Extra

We assume that all of the heat transfer in the calorimeter is from the water to the ice. Design an experiment to determine the heat transferred to the environment surrounding the calorimeter.

Experiment 59

Energy and Changes of State

Problem
What happens to the kinetic and potential energy of molecules during a change of state?

Introduction
You have learned that temperature is a measure of the random kinetic energy of the molecules in a system. This means that the molecules of a solid must have the same kinetic energy as the molecules of the liquid if both are at the same temperature. But if that's so, why aren't their states the same? Why aren't they either both solids or both liquids. The answer has to do with the potential energy of the molecules and with the forces pulling those molecules toward each other.

You can investigate the roles of kinetic and potential energy during a change of state by making a cooling curve, in which the substance is heated until it melts, then is allowed to cool below its freezing point. In this experiment, you will make two curves at once. You will begin by melting a pure solid in a test tube. This will be done by heating the tube in a beaker of water on a small hotplate. The substance, phenyl salicylate (known as *salol*) will melt by the time the water has started to visibly steam. Then you will remove the beaker and tube from the hotplate and use your CBL unit and two temperature probes to follow the cooling of both the substance in the tube and of the water in the bath. One probe will follow the temperature of the substance under study, while the other traces the changes in temperature of the water bath.

Prelaboratory Assignment
✓ Read the **Introduction** and **Procedure** before you begin.
✓ Answer the Prelaboratory Questions.

1. Consider two solids. One has a melting point of 40°C, while the other melts at 60°C. What does this tell you about the relative strengths of the intermolecular forces present in each one?

2. What happens to the molecules of a substance when it changes from liquid to solid? Is freezing of a substance an endothermic process or an exothermic one?

Materials
Apparatus
CBL, CBL2, LabPro or similar interface
Graphing calculator with data collection
    software installed (DataMate for
    CBL2 or LabPro)
Stainless steel temperature probes (2)
Beaker, 150-250-mL for water bath
Glass stirring rod

Reagents
Phenyl salicylate in 18 × 150 mm tube

Hot plate or heating unit
Hot pad
Safety goggles
Lab apron

Safety

1. **Caution:** The hot plate will be hot, but looks the same as when it's cool. Be sure to use a hot pad to move the water bath or the sample test tube.
2. Although you will not be handling any reagents directly, always observe common sense practices when working with chemicals.
3. Wear safety goggles and a lab apron at all times in the laboratory.

Procedure

Original CBL Set-Up. Following the instructions for your graphing calculator, prepare the CBL to collect temperature readings (ChemBio for TI-83+; the TEMP program for other models). Connect one temperature probe to Channel 1, the other to Channel 2 of your CBL unit. Turn on the calculator and the CBL and be sure that the connecting cable between them is firmly in place.

Choose Set Up Probes from the Main Menu. For Number of Probes, select 2; [ENTER]. At the next screen, select TEMPERATURE; [ENTER]. You will be told to use the lowest number channels available; select Ch1; [ENTER]. The next screen will ask to what channel the other probe is connected: 2; [ENTER].

When the calculator returns you to the Main Menu, select option 2: Collect Data, followed by TIME GRAPH. You will be asked for Time Between Samples: 15 [ENTER]; then for the number of samples: 100 [ENTER]. The next screen will give you a summary of your selections; [ENTER]. You will then be asked if you want to keep the time set up; you do--[ENTER]. When the calculator asks you for Y Minimum and Y Maximum, select 25 and 50, respectively. For Yscl, enter 1.0. This will allow you to collect your data as a function of time and will give you a total run time of 25 minutes of data collection.

CBL2, LabPro Set-Up. Both of these interfaces will recognize the stainless steel temperature probe on their own. You need only set the MODE to TIME GRAPH, with intervals of 15 seconds and a total of 80 samples (20 minutes).

Note: You can perform step 1, below, while the hotplate is warming the beaker and tube.

1. A sketch of the apparatus is shown in **Figure 1**. Place the probe connected to Channel 1 gently into the tube with the salol; do not try to force in down into the solid. As the solid melts the probe will drop easily into the molten sample. When this starts to happen, use the probe to **gently** stir the solid sample. This will help it melt more quickly. Place the other probe in the beaker of water. It may help to drape the probe cable over a utility clamp to keep the probes from falling out of the tube.

   Caution: Protect your hand with a glove or mitt when moving a hot container.

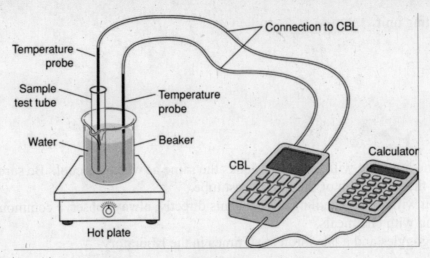

Figure 1
CBL set-up for changes of state

2. When all is ready, and the salol has melted, turn off the hotplate. Carefully lift the beaker with the tube from the hotplate and place it on the lab bench. Note the temperature reading on the calculator screen. When it has dropped below 50°, press 2-START on your calculator to begin data collection.

Salol has a tendency to exhibit a behavior known as *super-cooling*, in which its temperature will fall several degrees below the normal freezing point, then quickly rise back to the freezing point and level off there. This results in a V-shaped dip in the normal cooling curve. To help prevent super-cooling, gently use the probe to stir the contents of the tube. When the probe in the tube of salol no longer moves easily, stop stirring to avoid damaging the probe.

While one partner is attending to the tube and salol, the other should use a glass stirring rod to gently stir the water so that it cools at a steady rate.

3. At the conclusion of the run, return the beaker and tube to the hotplate. Heat the water enough that the salol melts again, so that you can remove the probe safely.

Cleaning Up

1. Clean the temperature probe by wiping gently with a cloth that has been dipped in alcohol. Blot the probe dry. If there appears to be some residual salol on the probe, clean it again.
2. Ask your teacher whether the water in the beaker should be poured down the sink or kept for another class.
3. Return the test tube containing the salol to the location specified by your teacher.
4. Wash your hands thoroughly before leaving the laboratory.

Analysis and Conclusions

Complete the **Analysis and Conclusions** section for this experiment either on your Report Sheet or in your lab report as directed by your teacher.

1. If they are available, use the program, Graphical Analysis™ and Graph-Link cables to transfer your data to a computer so that the graph can be printed and handed in with your report.

2. Describe the shapes of the two curves on your graph. Each shows time/temperature behavior, but one is for the water bath while the other is for the solid in the tube. On your printed graph, indicate which curve represents which substance. Include separate descriptions of the two graphs.

3. As the salol changed from liquid to solid, its temperature was constant.
   a. At what temperature did this occur? The temperature at which you observed the plateau is the freezing point of salol, as determined by your experiment.
   b. What can you say about the kinetic energy of the salol molecules along the temperature plateau?

4. During the time that the temperature of the salol remained constant, the temperature of the water in the beaker continued to fall. Was this a change in kinetic energy or potential energy for the water? Explain.

5. During the same time period, the salol temperature was constant, however it must have been losing energy also.
   a. Why must it have been losing energy?
   b. If its temperature is constant, what type of energy must it have been losing? Explain.

6. While a liquid is boiling its temperature remains constant even though you continue to apply heat. How must the water be storing the energy you are adding? (In other words, is the heat energy that you add being used to increase the kinetic energy or the potential energy of the molecules?)

7. On the basis of your answer to question 6, explain why you would receive a more severe burn from steam at 100°C than you would from (liquid) water at 100°C.

Experiment 60

Vapor Pressure of Water
An Experimental Determination

Problem

How can the vapor pressure of water be measured as a function of temperature?

Introduction

At very low temperatures (temperatures near the freezing point), the rate of evaporation of water (or any liquid) is negligible. But as its temperature increases, so does the rate at which the liquid evaporates. When a liquid evaporates in a closed container, the molecules of vapor exert pressure on the container walls, as any gas does. The pressure exerted in a closed container by an evaporating liquid is known as its *vapor pressure*. In this experiment you will measure how the vapor pressure of water changes with temperature.

Prelaboratory Assignment

✓ Read the **Introduction** and **Procedure** before you begin.

✓ Answer the Prelaboratory Questions.

1. Write the ideal gas law
   a. In standard form
   b. Rearranged to solve for n, in terms of P, V, and T
   c. Rearranged to solve for P, in terms of n, V, and T.

2. At any given temperature, what gas other than the trapped air is present in the inverted graduated cylinder?

3. State Dalton's Law of Partial Pressures. Explain how it applies to this experimental procedure.

Materials

Apparatus
10-mL graduated cylinder
1-L beaker, preferably tall-form
Thermometer
Large tray or sink for catching overflow
Hot plate or heating unit
Hot pad
Safety goggles
Lab apron

Reagents
Tap water
Ice

Safety

1. Safety goggles and a lab apron must be worn at all times in the laboratory.
2. Be careful using the hotplate; it is very hot, although it will not appear so. Use a hot pad to protect your hand when moving the water bath or hot glassware.
3. If you are using a mercury thermometer; be very careful. Mercury vapor is very poisonous. If the thermometer breaks, notify your teacher right away.

Procedure

If you are not using a Report Sheet for this experiment, create a Summary Table which includes the following headings: Observed Volume; Corrected Volume; Temperature (K); Mol Trapped Air; Barometric pressure; Pressure of air; Pressure of H_2O

1. Place 7-8 mL of water in a 10-mL graduated cylinder, then invert the graduated cylinder in a tall-form one liter beaker or similar tall, heat-tolerant container of water. **Note:** The entire graduated cylinder must be under water. Heat the system on a hot plate or ring-stand to a temperature of about 80°C, with gentle stirring to ensure even heating. **Do not use the thermometer to stir.**

2. Remove the beaker and graduated cylinder from the heat source, read the volume of moist air in the cylinder to the nearest 0.1 mL and record your value in the Summary Table. **Note:** You may have to lift the graduate above the water level in the beaker to make this and later volume readings. This should not affect your results so long as you work quickly. Be sure you do not lift the open end of the graduated cylinder above the water level.

3. Place the beaker and graduated cylinder in a tray or similar container that can catch overflow. Then add ice to the beaker, one or two pieces at a time, stirring gently to maintain a uniform temperature throughout the system. Continue the gradual cooling, recording the volume of moist air in the graduated cylinder at 5-degree intervals, down to a temperature of 40°C. Record these volumes and temperatures in the Summary Table.

4. Once the temperature has dropped below 40°C, you are ready to get a final reading. Add a handful of ice to the beaker of water and stir gently. If this is not sufficient to cool the water below 5°C, add more ice. When the temperature has fallen below 5°C, take a final reading of the volume and temperature.

5. Record the barometric pressure (torr).

Cleaning Up

1. All water used in the experiment may be poured down the sink, but it might be better used to water plants or trees.
2. Return all glassware and the thermometer to the proper locations.
3. Wash your hands before leaving the laboratory.

Analysis and Conclusions

Complete the **Analysis and Conclusions** section for this experiment either on your Report Sheet or in your lab report as directed by your teacher.

1. The graduated cylinder was calibrated to be read in an upright position. To allow for the fact that your readings were made with the cylinder inverted, the volume should be corrected by subtracting 0.2 mL from each recorded value. Record the corrected volumes in the Summary Table.

2. Below 5°C the vapor pressure of the water is negligible so at this point we assume that the only gas in the graduated cylinder is the trapped air. We can further assume that the pressure in the cylinder is equal to the barometric pressure. Use these assumptions and the Ideal Gas Law to calculate the number of moles, *n*, of air in the graduated cylinder.

3. Use your volume measurements for each of the other temperatures and the value of *n* obtained in question **2** to calculate the pressure due to air at each of the other temperatures. Answers will be *less than* the barometric pressure, since at each of these temperatures the graduated cylinder contains both air and water vapor.

4. At each of the temperatures, the difference between the calculated pressure (from question **3**) and the barometric pressure is the vapor pressure of water at that temperature. For each temperature enter the atmospheric pressure, the pressure due to air, and the vapor pressure of water in the appropriate columns of your table.

5. Plot vapor pressure as a function of temperature on graph paper. Draw the best smooth-curve fit you can.

6. Compare your results with the accepted values and curve shape, as found in your text or other references. Discuss your findings.

Experiment 61

Solution Properties

Problem

How does a particular substance dissolve in different solvents? How can we choose the best solvent for a substance? What factors affect the rate at which a substance dissolves?

Introduction

A **solution** is a system in which one substance is dissolved in another. Each solution has two components, the **solvent** which is the "dissolver" and the **solute** which is the substance being dissolved. The solvent is always present in larger amount than the solute. Water is the most common solvent in a beginning chemistry laboratory and in our everyday lives. If salt (sodium chloride) is dissolved in water, the solution is called a sodium chloride solution. In this case water is the solvent and sodium chloride is the solute.

Solutions are mixtures. This means that a solution can have a variety of compositions. More solute can be dissolved or less solute can be dissolved. Solutions are homogeneous mixtures since the solute remains mixed in the solvent.

When a solute and a solvent are mixed several possibilities exist:

1. The solute can dissolve completely into the solvent to form a homogenous mixture. In this case we say that the solute is *very soluble*.
2. The solute can dissolve a little bit into the solvent with some remaining undissolved. In this case we say the solute is *slightly soluble*.
3. The solute may not dissolve at all into the solvent. Now we have a heterogeneous mixture. If the solute is a solid it is said to be *insoluble*. If the solvent and solute are both liquids they are called *immiscible*.

Chemists also use other terms to indicate how much solute has dissolved in a solution. These terms all indicate the relative **concentration** of the solution. One qualitative system for indicating concentration is to call a solution with relatively little solute *dilute*, while a solution with a relatively large amount of solute dissolved is *concentrated*. If we want to compare solution concentration to the maximum amount of solute, which will dissolve at a particular temperature, we can use a different system. Here, a solution that contains less than the maximum amount of solute is called *unsaturated*, while a solution containing the maximum amount of solute possible at that temperature is a *saturated* solution. It is even possible to make a solution that contains more than the normal maximum amount of solute at a particular temperature. In this case the solution is called a *supersaturated* solution.

Prelaboratory Assignment

✓ Read the **Introduction** and **Procedure** before you begin.
✓ Answer the Prelaboratory Questions.

  1. How can you determine which substance is the solvent and which is the solute in a solution?

  2. How can you determine if two liquids are immiscible?

  3. What is the difference between a saturated and an unsaturated solution?

Materials

Apparatus
Test tubes (18 x 150)
Test tube rack
Spatula
2 small beakers (100 –150-mL)
Stoppers for the test tubes
Heating unit or hot plate
Weighing paper
Safety goggles
Lab apron

Reagents
Petroleum ether
Iodine water
Kerosene
Isopropyl alcohol
Rock salt (coarse sodium chloride)
Table salt (fine sodium chloride)
Ammonium chloride

Safety

1. Always wear safety goggles and a lab apron in the laboratory.
2. Do not pour solvents other than water down the drain.
3. Be careful when using the heating unit. Remember that hot and cold objects look alike. Protect your hand when transferring hot objects to prevent burns.
4. Several of the liquid reagents are flammable. Keep all liquids (other than water) away from flames.
5. When you shake a stoppered test tube, be sure to keep your thumb or forefinger on the stopper, to keep it from flying off.

Procedure

Part 1   Solubility

1. Place about 2 mL of petroleum ether in a test tube. Add about 5 mL of water to the same test tube. Record the position of each substance in the test tube. Stopper the test tube and shake (10 s). Unstopper the test tube and place it in the rack. Wait 2 minutes and record your observations.

2. Add 5 mL saturated iodine water to the test tube. Record the color of each layer. Stopper and shake for 20 seconds. Remove the stopper and wait 2 minutes. Record your observations.

3. Dispose of the test tube contents in the waste container provided by your teacher.

4. You will need 3 dry test tubes for this step. Place the following into each tube:

Tube #1	1 mL isopropyl alcohol	1 mL kerosene
Tube #2	1 mL water	1 mL kerosene
Tube #3	1 mL water	1 mL isopropyl alcohol

Stopper each tube and mix by shaking gently for 10 seconds. Record your observation (be sure to indicate which ones are miscible)

Part 2   Rate of Dissolving

1. You will need 2 dry test tubes for this step. Place the following into each tube:

Tube #1	0.5 cm table salt	10 mL water
Tube #2	0.5 cm rock salt	10 mL water

Stopper both tubes and shake both at the same time, recoding the number of seconds required to completely dissolve the salt.

2. Place 50 mL water in each of two small beakers. Heat the water in one of the beakers to boiling. Allow it to cool 1 minute.

3. Measure two samples of 0.5 g table salt. Add a sample of table salt to each beaker and record the time needed to dissolve the salt in the hot water.

4. As soon as the crystals have dissolved in the hot water carefully pick up the beaker and tilt it back and forth. Observe the layer of salt solution at the bottom of the beaker. Do the same with the other beaker.

5. You will need 4 dry test tubes for this step. Place the following into each tube:

Tube #1	1.0 g NaCl	5 mL water
Tube #2	1.4 g NaCl	
Tube #3	1.0 g NH_4Cl	5 mL water
Tube #4	1.4 g NH_4Cl	

Stopper tubes 1 and 3 and shake until the salts are completely dissolved.

6. Add the contents of test tube #1 to test tube #2 and add the contents of test tube #3 to test tube #4. Stopper and shake both tubes for 3 minutes. Record your observations.

7. Place both *unstoppered* test tubes into a beaker of boiling water. Agitate the tubes occasionally as you heat them for 5 minutes. Record your observations.

8. Place both test tubes in a beaker of cool water for about 1 minute. Remove and place in test tube rack. Observe over the next several minutes.

Cleaning Up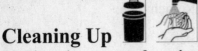

1. Place the contents from the test tubes in Part 1 in the waste containers provided by your teacher.
2. The contents of the test tubes in Part 2 can be put down the drain with plenty of water.
3. Wash your test tubes thoroughly and rinse well with distilled water.
4. Wash your hands thoroughly before leaving the laboratory.

Analysis and Conclusions

Complete the **Analysis and Conclusions** section for this experiment either on your Report Sheet or in your lab report as directed by your teacher.

Part 1   Solubility

1. Which liquid is more dense, petroleum ether or water? Support your answer with evidence.

2. Are petroleum ether and water miscible? Support your answer with evidence.

3. Is iodine more soluble in water or petroleum ether? How can you tell?

4. Which of the pairs of liquids in step 4 are miscible?

Part 2   Rate of Dissolving

1. Did the coarse salt dissolve faster or slower than the fine salt?

2. Did the salt dissolve faster in hot or cold water?

3. If salt is soluble in water, why could you see a layer of dissolved salt at the bottom of the beakers in step 4?

4. Which of your test tubes in step 5 containing 1.0 g salt per 5 mL water were unsaturated?

5. Which test tube(s) containing 2.4 g salt per 5 mL water are saturated at room temperature?

6. Which salt is least soluble at higher temperature?

7. Classify the salt solutions at the higher temperature as unsaturated or saturated.

8. Did the salt solutions which were unsaturated at higher temperature become saturated as they cooled? Give evidence to support your answer.

9. If two liquids in a container are immiscible how can you tell which is which?

10. What do you conclude about particle size and rate of dissolving from your experiment in step 1?

11. What do you conclude about temperature and dissolving from your experiment in step 4?

12. Write a paragraph summarizing your conclusions regarding temperature and unsaturation.

Something Extra

Suppose you were given a mixture containing sodium chloride and benzoic acid, both fine white crystals. Design a procedure to allow you to separate the two. Benzoic acid is soluble in hot water but not in cold water.

Experiment 62

Polar and Nonpolar Solvents

Problem

How does the molecular polarity of a liquid affect its ability to act as a solvent?

Introduction

As you know, some substances dissolve very well in water, some only sparingly, and others not at all. Why? The answer lies in the polarity of the molecules in both the solvent and the solute. You encountered the concept of molecular polarity in Chapter 12. As you learned there, a molecule is polar if its center of positive charge is in a different location than its center of negative charge. This, in turn, is determined by the arrangement and polarity of the individual bonds within the molecule. Water is perhaps the most important example, so let's consider it. As shown in the sketch, the water molecule consists of an oxygen atom with two hydrogen atoms bonded to it. If you were to draw lines connecting the nuclei of the two hydrogen atoms to the nucleus of the hydrogen atom, you would find that the angle they describe is about 105°.

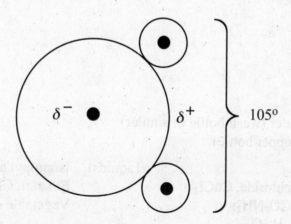

Further, because the electronegativity of hydrogen is only 2.1, while that of oxygen is 3.5, both O-H bonds are very polar. This means that the center of negative charge (represented by δ-) is close to the oxygen atom, while the center of positive charge (δ+) lies midway between the hydrogen atoms (review Chapter 12 for a more in-depth discussion).

The charge separation in water is responsible for most of its unique properties. In this case, however, our focus is on how the polarity of water can be used to explain its ability to dissolve many different substances; it also explains why certain types of materials do not dissolve well in water.

On the other hand some solvents contain nonpolar molecules and have very different behavior from water. We will test one such nonpolar solvent, hexane, in this experiment.

Prelaboratory Assignment

✓ Read the **Introduction** and **Procedure** before you begin.
✓ Answer the Prelaboratory Questions.

1. Why is it so important to thoroughly clean and dry the tubes before beginning Step 6?

2. Based on your previous experiences with copper(II) chloride and other compounds containing the copper(II) ion, how will you be able to tell whether $CuCl_2$ has dissolved or not (other than by the disappearance of crystals)?

3. Copper(II) chloride and ammonium nitrate are ionic compounds. Urea and isopropyl alcohol are weakly polar, while vegetable oil and naphthalene are completely nonpolar. Prepare a table predicting the solubility of each of the eight solutes in each of the two solvents, water and hexane.

4. Make a sketch showing the attractions of water molecules for each other. Indicate the attractive forces by dashed lines.

5. Make separate sketches indicating the attractive forces between water molecules and
   a. anions   b. cations.

Materials

Apparatus

8 calibrated test tubes, or 8 small tubes and a wax pencil or permanent marker
Corks or stoppers to fit tubes
Metric ruler
Wash bottle
Safety goggles
Lab apron

Reagents

Solvents: Distilled water (wash bottle or similar)
               Hexane (dropper bottle)

Solutes: (Solids)   (Liquids) Isopropyl alcohol, CH_3-CH(OH)-CH_3
               Copper(II) chloride, $CuCl_2$ Ethanol, CH_3CH_2OH
               Urea, H_2N-CO-NH_2 Vegetable oil
               Sucrose, $C_{12}H_{22}O_{11}$
               Ammonium nitrate, NH_4NO_3
               Naphthalene, $C_{10}H_8$

Safety

1. Naphthalene has a relatively strong odor; keep its container tightly closed when not in use.
2. Many of the reagents are toxic. Avoid ingestion and contact. Wash your hands thoroughly before leaving the laboratory.
3. Wear safety goggles and a lab apron at all times in the laboratory.

Procedure

1. If the tubes are not already calibrated, use a ruler and a wax pencil or permanent marker to mark a depth of 1 cm on each of seven test tubes.

2. Fill each tube to the 1-cm mark with distilled water from the wash bottle.

3. Add one of the solutes to each of your tubes.
    - ✓ **For solid solutes** use a soda straw scoop to place *a few granules* of the solute in the tube. This is a case where less is better. If you add too much of a solid, you may not be able to tell whether any dissolves.
    - ✓ **For liquid solutes** add about 8-10 drops from the dropper bottle to the water in the tube.

4. Mix the contents of the tubes by holding the top of the tube and flicking the side with your finger. Reserve final judgment on whether the solutes dissolve or not until the tubes have had a few minutes to stand and the you have mixed the contents two or three times. Some solutes will dissolve but do so rather slowly. Record your observations in a data table.

5. Wash the tubes and *dry them thoroughly.* Important; be sure the tubes are completely dry. If necessary, remark the 1-cm depth on the tubes.

6. Fill each tube to the 1-cm mark with hexane from the dropper bottle. Repeat Steps 3 and 4, testing each of the seven solutes for the ability to dissolve in hexane. Record your observations as before.

7. When all your hexane tests are complete, add a few drops of water to one of the tubes containing a hexane solution. Observe what happens to the water; record the result.

Cleaning Up

As you have just seen, hexane and water don't mix, so normal washing methods are not going to work. It is also not appropriate to pour organic solvents down the drain, so special methods are needed.

1. Empty the tubes containing the hexane mixtures into an organic waste container, as directed by your teacher.
2. Rinse each tube three or four times with about 1 mL of hexane at a time. At this point, the tubes should have nothing but residual hexane in them. The hexane will evaporate to dryness in a short time. Consult your teacher to see whether the tubes are to be allowed to air-dry on their own, or if you are to dry them manually.
3. Return all equipment to its proper location.
4. Wash your hands thoroughly before leaving the laboratory.

Analysis and Conclusions

Complete the **Analysis and Conclusions** section for this experiment either on your Report Sheet or in your lab report as directed by your teacher.

1. Divide the solutes into two or three groups, as follows. One group should include any solutes that dissolve only in water. The second group is for those solutes that dissolve only in hexane. If any solutes showed at least some solubility in both hexane and water, they constitute a third group.

2. What similarities can you find for those solutes in your first group (water-soluble only)? In particular, look for structural characteristics.

3. Look up the structures of the solutes in your second group (hexane-soluble). If they are available, examine models of these molecules. Use the molecular shapes, and the arrangement and electronegativities of the atoms in the molecules to help you explain their solvent preference.

4. Finally, for those solutes that had some affinity for both water and hexane:
   a. Try to identify the structural characteristics that attract them to water. Sketch the molecules, and indicate the water attracting regions on your sketch.
   b. Now try to identify the part(s) of the same molecules that make them soluble in hexane. As with question 3, molecular models may make your task much easier. Sketch the molecules and indicate the hexane attracting (or water repelling) regions on your sketch.

Experiment 63

Temperature and Solubility

Problem

How does temperature affect the water solubility of an ionic solid?

Introduction

When the proportion of solute to solvent reaches the point that no more of the solute can be dissolved into solution, the solution is said to be *saturated*. Some salts, such as NaCl, KCl, and NH_4Cl, are very soluble in water. For example, about 36 g of NaCl will dissolve in 100 g of water at room temperature (25°C). Other salts are far less soluble; silver chloride, AgCl, reaches saturation when only about 2×10^{-4} g is dissolved per 100 g of water.

Solubility is usually measured and reported in one of two ways. One is the mass of solute that will dissolve in 100 grams of water (or other solvent); this is the one we will use. The other method, known as the molar solubility, is the number of moles of the salt that dissolves in one liter of solution.

Some solutes, including potassium nitrate, have a tendency to remain in solution even after being cooled to well below the saturation point. This phenomenon is known as *supersaturation*.

In this experiment you will explore the effect that a change in temperature has on the solubility of a typical ionic salt. You will study the solubility of potassium nitrate, KNO_3, in water by adding different quantities of the salt to a fixed (constant) mass of water at a temperature close to the boiling point of the solvent. Each solution will be allowed to cool, and the temperature at which crystals begin to appear will be noted; this is the signal that the system has cooled to the saturation point. Data for the various combinations will be collected and a plot of concentration vs. saturation temperature will be prepared.

Prelaboratory Assignment

✓ Read the **Introduction** and **Procedure** before you begin.
✓ Answer the Prelaboratory Questions.

1. Write the complete ionic equation for dissolving potassium nitrate in water.
2. Make a molecular level sketch of the potassium nitrate solution. Include a representation of the interactions of the water molecules and the ions. Name and describe the forces at work in this system

Materials

Apparatus
4 test tubes, 75- or 100-mm
Thermometer
Balance, milligram or centigram precision
100-mL beaker (or similar)
Test tube clamp
Hot plate **or** ring stand, burner, and wire gauze
Safety goggles
Lab apron

Reagents
$KNO_3(s)$, about 4 g
Distilled water

Safety

1. Safety goggles and a lab apron must be worn at all times in the laboratory.
2. **Avoid burns;** work carefully around flames or hot plates. Handle the hot test tubes only with the test tube clamp, not with your fingers.

Procedure

If you are not using a Report Sheet for this experiment design a Data Table with the following column headings:

Test tube	(1) Mass of KNO₃	(2) Mass of H₂O	(3) g KNO₃ / g H₂O	(4) Crystallization Temp (°C)	(5) Concentration of solution (g/100g H₂O)

1. Prepare a hot water bath by filling your 100-mL beaker two-thirds full with water and placing it on a hot plate or ring stand with wire gauze.

2. Label four test tubes A, B, C, and D, and add the following quantities of potassium nitrate to each. Note that the amounts are given as ranges; for this experiment, we will round masses to the nearest centigram. In column (1) of the Data Table, record the mass of the empty tube and the mass of tube and solid.

Tube	Mass of KNO₃(s)
A	0.35 - 0.45 g
B	0.75 - 0.85 g
C	1.10 - 1.20 g
D	1.50 - 1.60 g

Note: Place a small beaker on the balance pan to support the test tubes when you weigh them. If you are using a top-loading electronic balance, you can zero the balance with the empty beaker in place, then place the tube in the beaker, and record the total mass.

3. Use a clean thin-stem pipet to add 20 drops of distilled water to each tube. Weigh the tubes again and determine the mass of water in each. If the mass increase due to the addition of water does not fall in the range of 0.80 - 1.10 g, add a few more drops. Record the mass of water in each tube in column (2) of the data table, then use columns (1) and (2) to calculate the ratio of mass of solute to mass of water for column (3).

   Note: Remove the tube with the solid from the balance before adding the 20 drops of water. This is to protect the balance from possible spills.

4. Place all four labeled tubes in the water bath with your thermometer in beaker D. Stir the mixture in D gently, using an up-and-down motion of the thermometer, until the solid in that tube dissolves completely. At this point, the solid in the other tubes will likely also be completely dissolved, even without stirring.

5. Remove the beaker from the heat source to a safe place. Use your test tube clamp to remove test tube D (only) from the beaker and allow the tube and contents to cool as you watch for the first sign of crystallization.

   Notes: 1. Move the thermometer gently up and down to ensure constant, even cooling and to encourage crystal formation. Recall that potassium nitrate has a tendency to form supersaturated solutions.
   2. You are looking for the formation of colorless crystals in a colorless solution; they will not be easy to detect, so you must watch closely.

   The instant crystallization begins, record the temperature in column (4) of the Data Table, then remove the thermometer from the tube, wipe it once with a clean paper towel to remove any potassium nitrate crystals, warm it briefly in the hot water in the beaker, wipe again, then transfer it to tube C.

6. Repeat Step 5 with tube C, then clean the thermometer and transfer it to tube B, and so on until all four tubes are done.

Cleaning Up

1. While potassium nitrate does not pose a serious environmental problem, it can be reused for this and other experiments. Your teacher will provide a container for collection of the water-KNO_3 mixtures.
2. Rinse the tubes and beaker well with water and return all equipment to its proper location.
3. Wash your hands thoroughly before leaving the laboratory.

Analysis and Conclusions

Complete the **Analysis and Conclusions** section for this experiment either on your Report Sheet or in your lab report as directed by your teacher.

1. Use the data from column (3) of the data table to determine the concentration of each saturated solution in grams of solute per 100 grams of water. Show your calculation for tube A. Record all concentrations in column (5) of the table.

2. Plot a graph of solubility of KNO_3 in grams/100.0 g of water vs. temperature (temperature is the horizontal axis) . Label the axes of your graph and give it an appropriate title. Draw the best smooth-curve through the points, extending your graph from 0°C to 100°C.

3. **Using your graph**, determine the solubility of KNO_3, in g/100 g H_2O, at:
   a. 30°C   **b.** 70°C   **c.** 0°C   **d.** 90°C

4. **From your graph**, determine the temperature at which each of the following mixtures of potassium nitrate in water would be saturated solutions.
   Note: Use your graph; do not calculate the values.

   a. 45 g KNO_3 in 100. g H_2O **b.** 20 g KNO_3 in 100. g H_2O
   c. 25 g KNO_3 in 25 g H_2O **d.** 100 g KNO_3 in 250. g H_2O

5. Define the terms, *saturated, unsaturated,* and *supersaturated* as they apply to solutions. Use complete sentences.

6. Classify the following solutions as saturated, unsaturated, or supersaturated, based on your graph. Defend your answers. Note that **c.** and **d.** require you to convert from the concentrations given to g/100 g H_2O before you refer to your graph.
   a. 75 g KNO_3/100. g H_2O at 40°C   **b.** 60 g KNO_3/100. g H_2O at 50°C
   c. 100 g KNO_3/75 g H_2O at 80°C   **d.** 175 g KNO_3/250. g H_2O at 40°C

Experiment 64

Chloride in Water

Problem

What is the chloride ion concentration in your water?

Introduction

Water samples, even bottled water, naturally contain chloride ions. In this lab you will determine the concentration of the chloride ions in various samples of water. You will be able to estimate the accuracy of your technique by testing solutions with known concentrations of chloride ions.

The technique you will use is called a titration. This technique involves adding a measured volume of a solution of known concentration to a measured volume of a solution of unknown concentration. The two solutions react with each other.

You will use aqueous silver nitrate to titrate your water sample. The silver ion reacts with chloride ion to form the white solid silver chloride. This solid is insoluble in water, so a cloudy solution indicates the presence of chloride ions in your water. The indicator for the titration is called dichlorofluorescein. The color change that indicates the end of the titration is from light green to pink. Because of the presence of silver chloride, the solution will resemble strawberry milk.

Prelaboratory Assignment

✓ Read the entire experiment before you begin.
✓ Answer the Prelaboratory Questions.

1. Provide balanced net ionic equations for all chemical reactions in this lab.
2. What is the purpose of titrating solutions of potassium chloride with known concentrations?
3. What is the purpose of diluting the potassium chloride solutions and titrating them?
4. Provide calculations for all dilutions to be made in this lab.

Materials

Apparatus
Safety goggles
Lab apron
Graduated cylinder
Ring stand
250-mL beaker
Buret clamp
Buret

Reagents
Bottled water
Water sample
Dichlorofluorescein
Dextrin
Potassium chloride (5.00×10^{-3} M)
Silver nitrate (5.00×10^{-3} M)

Procedure

Part 1   Determining Accuracy of the Titration

1. Fill the buret with 5.00×10^{-3} M silver nitrate. Record the initial reading.

2. Use the graduated cylinder to measure about 10.0 mL of the 5.00×10^{-3} M potassium chloride solution and place it in the 250-mL beaker. Record the actual volume.

3. Add 2-3 drops of the dichlorofluorescein and a spatula-tip amount of dextrin. The dextrin is a coagulating agent that helps the silver chloride settle out of the solution. Swirl to dissolve.

4. Slowly add the silver nitrate solution until the solution turns pale pink. Record the final volume reading on the buret.

5. Make 50.0 mL of 5.00×10^{-4} M silver nitrate and 50.0 mL of 5.00×10^{-4} M potassium chloride solutions from the given 5.00×10^{-3} M solutions. Titrate the potassium chloride solution as before (use 10.0 mL of 5.00×10^{-4} M potassium chloride). Record the initial and final volume readings on the buret.

6. Make 50.0 mL of 5.00×10^{-5} M potassium chloride solution from the 5.00×10^{-4} M solution and titrate 30.0 mL of this solution with the 5.00×10^{-4} M silver nitrate solution. Record the initial and final volume readings on the buret.

Part 2   Testing Water Samples

1. Bring in a sample of water from your home faucet or from a local lake, river, or stream.

2. Titrate 10.0 mL of your water sample twice. Once with 5.00×10^{-3} M silver nitrate solution and once with 5.00×10^{-4} M silver nitrate solution. Record the initial and final volume readings on the buret for each titration.

3. Titrate two 10.0 mL samples of water from the nearest drinking fountain in your school. Do one titration with 5.00×10^{-3} M silver nitrate solution and the second with 5.00×10^{-4} M silver nitrate solution. Record the initial and final volume readings on the buret for each titration.

4. Titrate two 10.0 mL samples of bottled water. Do one titration with 5.00×10^{-3} M silver nitrate solution and the second with 5.00×10^{-4} M silver nitrate solution. Record the initial and final volume readings on the buret for each titration.

Cleaning Up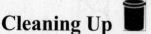

1. Empty the buret and rinse thoroughly. Clamp it back in the buret clamp upside down to drain.
2. Dispose of all chemicals as instructed by your teacher.
3. Wash your hands thoroughly before leaving the laboratory.

Analysis and Conclusions

Complete the **Analysis and Conclusions** section for this experiment either on your Report Sheet or in your lab report as directed by your teacher.

1. Calculate the chloride ion concentration for the home sample, school sample, and bottled water sample

2. Determine the percent error for each of the three titrations in Part 1 by comparing the known chloride concentrations to those you found by titration.

3. Which known concentration of potassium chloride gave the smallest percent error? Was this expected? Explain.

4. Compare your school sample and bottled water results with your classmates. Are they similar?

Something Extra

Contact your local water company and the bottled water distributor and obtain data on the chloride ion concentrations. Do these agree with your results?

Experiment 65

Hard Water Analysis

Problem

How can we determine the total hardness in tap water?

Introduction

If you let tap water boil for long periods of time, it begins to leave a film on the walls of the container. You may have seen this sort of film on beakers and flasks in the lab, or in the pots and pans in your kitchen at home. Most of these whitish deposits are residues of the carbonate and sulfate salts of calcium, usually with small amounts of the sulfates and carbonates of magnesium mixed in as well. The presence of these minerals in water is the condition which we call "hard" water, and the more minerals you have, the harder your water is. The tendency of ordinary soaps to form precipitates with the Ca^{2+} and Mg^{2+} in hard water results in a grayish film being left on clothes, and accounts for the widespread use of detergents in today's world.

In this experiment you will explore one type of test which can be run on ordinary tap water to determine its degree of hardness. The analysis involves a technique known as a "titration". The metal ions (Ca^{2+} and Mg^{2+}) in hard water have a particularly strong attraction for a large organic molecule called ethylenediamine tetraacetic acid, or simply EDTA. The metal ions and the EDTA molecules combine to form a large complex ion, known as a "chelate", in which the EDTA surrounds and traps the metal cation. (Chelates are important molecules biologically. Hemoglobin is another example of a chelate; it holds the iron in your blood.)

In this titration, you will carefully add a solution of EDTA to samples of tap water, stopping when all of the metal ions have been chelated. To help you identify the endpoint, you will add a few drops of an *indicator*--a dye which changes color when all of the metal ions are gone. Since the reaction only works in a moderately alkaline medium, it will be necessary for you to add a solution known as a buffer to your samples; a buffer maintains the pH (acidity-alkalinity) of a system at the desired level. This will ensure that the endpoint is clear and unobscured.

Because there is generally far more calcium than magnesium in water samples, the results are ordinarily reported in parts per million (ppm) of Ca. We will follow that convention here.

Prelaboratory Assignment

✓ Read the **Introduction** and **Procedure** before you begin.
✓ Answer the Prelaboratory Questions.

1. Water softeners remove cations such as Ca^{2+} and Mg^{2+} from the water, but they must also replace these cations with less troublesome cations such as Na^+. Why?

2. One of your four beakers will serve as a control. What does that mean?

3. The final color of the titration vessels (A-C) is often a lighter shade of blue than the control. Why?

4. Notice that steps 2 and 3 direct you to put tap water in beakers A-C, *then* put distilled water in beaker D. Suggest a reason why it would not be good to fill the control beaker first, then fill the titration vessels.

Materials

Apparatus
4 beakers, 20-mL to 50-mL
10-mL graduated cylinder
Safety goggles
Lab apron

Reagents
Tap water (or artificially hardened water)
Buffer solution, pH 10, in dropper bottle
Distilled (or deionized) water, in wash bottle
0.010 M EDTA solution, in microtip pipette
Calmagite indicator solution, in thinstem pipette

Safety

1. Safety goggles and a lab apron must be worn in the laboratory at all times.
2. While none of the reagents involved has a known toxicity, safe laboratory practice is always expected.

Procedure

Part 1:   The Analysis

1. Label four small beakers, A-D. Beaker D will be the control; the other three will be used for your sample analyses.

2. Rinse a 10-ml graduated cylinder with tap water and shake it once to remove most of the residual drops. Do not dry the inside with towels. Now using the graduated cylinder pour exactly 5.00 mL of tap water into each of the beakers, A-C. Shake the graduate gently to be sure that the transfer from the cylinder to the beaker is as complete as possible.

3. Rinse your graduate thoroughly with distilled (or deionized) water, and shake as before to remove excess. Now put exactly 5.00 ml of distilled water in D, your control.

4. Add 2.0 ml of the pH 10 buffer, and 5 drops of calmagite indicator to each of your four beakers. Swirl gently to mix.

5. Using a micropipet add 1 drop of 0.010 *M* EDTA solution to beaker D, then swirl to mix. This should result in a clear, blue color. If it does not, continue adding EDTA dropwise, swirling after each addition, until the clear blue color appears. Record the number of drops needed. Set this control beaker aside for comparison with your titration samples.

6. Add 10 drops of EDTA to the mixture in beaker A, and swirl to mix. If the indicator does not change from red to blue, add 5 more drops of EDTA, and swirl again. Continue in this fashion, using 5-drop increments (keeping written count of the number of drops used), until you get a blue color that does not change back to red. Record the number of drops used. The change is a gradual one, and could require more than 100 drops, if the water being tested is quite hard. Refer frequently to the color of the control (beaker D). Remember that you want a pure blue color, with no trace of red or purple, and that the blue color may not be as dark as the color of the control.

7. Titrate samples B and C in the same way *except* that as you near the expected endpoint, you should add only one or two drops at a time, with swirling. This will permit you to get a more precise value of the number of drops required. If the results for your samples vary greatly, your teacher may recommend that you do one or more additional samples; your results will be judged in part on the accuracy and precision that you achieve.

Part 2:   Calibration of the EDTA Pipet
1. After you have completed all of your titrations, calibrate your pipet as follows:
   a. Put about 2 ml of water in your 10-ml graduated cylinder. Record the actual volume to the nearest 0.01 ml.
   b. Use your EDTA pipet to add exactly 20 drops of liquid to the water in the cylinder. Let the drops fall to the water surface; if they hit the sides, they may or may not get into the water in the bottom. Record the new volume of water in the cylinder to the nearest 0.01 ml. Note that this requires estimating between lines.

Cleaning Up
1. Rinse the contents of all four beakers down the drain with large amounts of water.
2. Return the EDTA and Calmagite pipets to their proper locations.
3. Make sure your work area is clean, and that all materials are put away
4. Wash your hands before leaving the laboratory.

Analysis and Conclusions
Complete the **Analysis and Conclusions** section for this experiment either on your Report Sheet or in your lab report as directed by your teacher.

1. Use your calibration data from Part 2 to determine the number of drops that is equivalent to 1.00 mL. Show your calculations.

2. Convert the number of drops you used in each titration to milliliters.

3. Calculate the number of moles of EDTA used in each titration, using the volumes you calculated in the previous question and the molar concentration of the EDTA, 0.010M. Show a sample calculation.

4. Given that one EDTA reacts with one calcium ion, calculate the number of grams of calcium that were present in each of your 5.00 ml samples. Show a sample calculation for Beaker A.

5. Assuming the density of water to be 1.0 g/mL (or 1000 g/L), a calcium ion concentration of 1.0 g/L would correspond to 1.0 g Ca^{2+}/1000 g water or 1000 g Ca^{2+}/ 1,000,000 g water or 1000 ppm (parts per million). Convert the results of question 4 from grams of Ca^{2+} per 5.00 g of H_2O to grams of Ca^{2+} per million grams of water (ppm). Show your calculation for Beaker A. Report both the individual sample values and the average value for the three trials.

6. Identify two major sources of experimental error in this analysis and suggest ways the procedure could be modified to minimize those errors.

Something Extra

1. If you have (or if someone you know has) a water softener in the home, *and* if you can do so without seriously disrupting the operation of the system, try taking samples for analysis (about 40-50 mL) of both the softened and the unsoftened water. A good way to do this is to draw your "unsoftened" sample from an outdoor tap-this water is usually not run through the water softener. Analyze at least three samples of each; discuss the significance of your results. It might be interesting to do this just before and just after the softener has been 'recharged' with salt.

2. If your kitchen (or other) faucet has a purification filter, try analyzing both the filtered and the unfiltered water. Discuss your results as outlined in **1, above**.

3. Repeat the experiment as described in the Procedure, but use 2 mL of water and 1 mL of buffer, instead of 5 mL and 2 mL, respectively. Compare both the hardness values obtained and the consistency (precision) of data, and discuss your findings.

4. Try substituting volumes of 10 mL for the water and 4 mL of buffer in the original procedure. Discuss as suggested in **3, above**.

Experiment 66

Acids and Bases

Problem

What properties distinguish acids from bases?

Introduction

There is a key difference between this experiment and any of the others that you have done. The difference is that it will be up to you to design much of the procedure. First, let's review some of the key concepts, covered in chapters 15 and 16.

Acids are compounds that increase the amount of hydrogen ion, H^+, in aqueous solution. Although you will not demonstrate this for yourself, a key feature of acids (and the reason for their name!) is that they have a sour taste. You will investigate some of the other properties common to acids in the course of your experiment.

Bases increase the amount of hydroxide ion in aqueous solutions. A characteristic of bases which you will *not* test for yourself is that they taste bitter. As with acids, you will look into some of the other characteristics of bases as part of this investigation.

In Chapter 15, you learned about solution concentration. The concentration of a solution refers to the amount of solute in a given amount of solution. In qualitative terms, we can say a solution is concentrated if it contains a high proportion of solute. The opposite of "concentrated" is dilute; a dilute solution contains relatively little of the solute per unit volume.

A more quantitative measure of concentration is molarity. The molarity of a solution is the number of moles of solute per liter of solution. Recall that moles represent numbers of particles, so molarity (also called "molar concentration") is a measure of the number of solute particles in each liter of solution. Thus, a solution that has a concentration of 1 M ("one molar") has half as many solute particles as the same volume of a solution that has a 2 M concentration. Two solutions that have equal concentrations must have the same number of solute molecules, even if the solutes are not the same.

The concept of molarity will be used extensively in calculations in later experiments. All you need for now is an understanding of the meaning of concentration: if two solutions have the same concentration, then equal volumes will contain equal numbers of solute particles. If the concentrations are different, the one with the higher concentration has more solute particles per unit volume than the more dilute one and the ratio of solute particles is the same as the ratio of molarities.

The second new concept is pH. You are probably familiar with pH from biology and you may have seen the words "pH balance" on shampoos and other cosmetic items. The pH is a measure of the relative acidity of a system. The lower the pH, the more acidic the system, so strong acids usually have pH values around 0 or 1. Values between 1 and 7 are common for very dilute acid solutions and for solutions of compounds called "weak acids." Solutions whose pH is above 7 are basic or alkaline. A solution that has a pH of 12 would be said to have a high alkalinity. A solution that is neither acidic nor alkaline is "neutral" and has a pH of 7.

An easy way to measure the pH of a system is with indicators. These indicators are dyes that change color depending on the pH (acidity or alkalinity) of a solution. You can either add a few drops of a solution of the indicator directly to the system you want to test, or you can touch a drop of the solution you're testing to a strip of paper that has been soaked in the dye.

Prelaboratory Assignment

✓ Read the **Introduction** and **Procedure** before you begin.

✓ Answer the Prelaboratory Questions.

1. Chocolate and coffee can have bitter tastes. With what class of compounds would you associate these two foods?

2. Cite at least two examples of foods (other than citrus fruit products) that you know must contain acids.

3. Based on your reading of Chapter 16, what is the net ionic reaction that occurs when any strong acid reacts with a strong base?

4. Explain the meaning of the word "strong" in relation to acids and bases.

Materials

Apparatus
24-well test plate
Plastic tooth picks
pH indicator paper
Glass stirring rod
Conductivity tester
Safety goggles
Lab apron

Liquid reagents will all be dispensed in thin-stem pipettes. Distilled water will be available in pipettes and wash bottles.

Reagents
0.1 M Acetic acid, $HC_2H_3O_2(aq)$ (a weak acid)
0.1 M $NH_3(aq)$ (a weak base)
Indicators, in microtip pipettes:
    congo red
    indigo carmine
    malachite green
    phenolphthalein
    universal indicator (a mixture of several other indicators)
1 M Hydrochloric acid, $HCl(aq)$
0.1 M Hydrochloric acid, $HCl(aq)$
0.1 M Nitric acid, $HNO_3(aq)$
0.1 M sodium hydroxide, $NaOH(aq)$
Distilled water
Magnesium, zinc, and copper (small pieces)

Safety

1. **Acids and bases are toxic. They are also corrosive to skin and clothing.** Wipe up all spills with large volumes of water. If either an acid or a base gets on your skin or clothing, rinse the affected area thoroughly for 5 minutes and notify the teacher.

2. Wear safety goggles and a lab apron at all times in the laboratory.

Procedure

You have eight questions to answer; it is up to you to decide how to find the answers to those questions. While the individual parts of the experiment do not have to be done in sequence, they have been arranged in a way that makes a logical progression. In the spirit of true research, each part is presented in the form of a question.

In each case, you are to design and carry out an experiment that will provide you with the information you need to answer the question. Following each question you will find such additional information you may need to help you formulate a plan of attack. For each of the questions posed in the Procedure, describe the process you followed and what observations you made. You will answer each of the questions in the **Analysis and Conclusions** section of your report.

1. **What happens when strong acids come into contact with active metals?**
   "Active," used in reference to an element, means it reacts readily with many other elements. Zinc and magnesium are considered active metals, while copper is relatively inactive.

2. **Are strong acids and bases strong electrolytes? How is conductivity related to concentration?** Recalling what electrolyte means, a quick look at the materials list should get you started. Feel free to use concentrations other than those provided to you. (**Hint:** What would be the concentration of a solution made by diluting one drop of the solution with 9 drops of water?)

3. **Is it possible to distinguish experimentally between strong and weak electrolytes on the basis of conductivity?** You're on your own for this one.

4. **If you add an indicator to a sample of a strong acid, what happens to the color of the indicator as you dilute the acid?** You may need to refill your distilled water pipette. Use a colored liquid indicator and try to be as quantitative as you can; if dilution does affect the color, what relationship can you find between the amount of water added and the change in color that you observe?

5. **How do indicators respond when you start with a fixed amount of a strong acid and add strong base to it, one drop at a time?** Acids and bases neutralize each other. Try following the process of neutralization, starting with 10-15 drops of acid, then adding a base of equal concentration 1 drop at a time. To check pH, touch a drop of the solution to a clean spot on the test strip and compare the color you get with the reference chart on the vial. One strip can be used for several tests. See how well you can identify the point at which the system becomes neutral. Repeat using one or more of the indicator solutions and compare the two methods. The name for the process you are investigating is **titration**. It is most often used as a means to determine the strength of an acid or base.

   The wells in your test plate can hold about 50 drops of the size that comes from a thin-stem pipet. Use enough acid that color changes will be gradual, but you also want to have enough room in the well so you can add an excess of the base.

6. **Do all indicators give the same results in an experiment such as the one you designed for Part 5?** Remember that Universal Indicator is a mixture; would it be reasonable to compare it with the others? What would you expect to happen with such a mixture?

7. **In what way (or ways) do weak acids and bases differ from the stronger ones in the various tests you've done here?** Time probably won't permit you to make a thorough inquiry of this question, so plan ahead.

8. **Is your school's tap water neutral?**

Cleaning Up

1. Use tweezers or forceps to remove remaining bits of metal. Rinse the metal pieces and then discard them in the solid waste container.
2. Acids and bases must be diluted before disposal. Since most of your experiments will involve dilution, it is safe to rinse the contents of the well plate in the sink as often as needed. It would be a good idea to rinse it with distilled water each time you clean the well plate. Use the wash bottle for this purpose.
3. Wash your hands thoroughly before leaving the laboratory.

Analysis and Conclusions

Complete the **Analysis and Conclusions** section for this experiment either on your Report Sheet or in your lab report as directed by your teacher.

Answer each of the eight questions, citing evidence from your observations to support your answers.
1. What happens when strong acids come into contact with active metals?

2. Are strong acids and bases strong electrolytes? How is conductivity related to concentration?

3. Is it possible to distinguish experimentally between strong and weak electrolytes on the basis of conductivity?

4. If you add an indicator to a sample of a strong acid, what happens to the color of the indicator as you dilute the acid?

5. How do indicators respond when you start with a fixed amount of a strong acid and add strong base to it, one drop at a time?

6. Do all indicators give the same results in an experiment such as the one you designed for Part 5?

7. In what way (or ways) do weak acids and bases differ from the stronger ones in the various tests you've done here?

8. Is your school's tap water neutral?

Experiment 67

Acid Rain

Problem
What is the normal range of pH for rainwater? Can soil neutralize the rainwater?

Introduction
At 25°C, a pH of 7.00 indicates a solution is neutral. Very few substances are neutral (pure water is, but pure water is not found in nature). Rainwater is naturally acidic due to dissolved carbon dioxide from the air.

In this lab you will determine a range of pH values of rainwater in your area. You will also determine the effect of soil on the acidity of rainwater.

Prelaboratory Assignment
✓ Read the entire experiment before you begin.
✓ Answer the Prelaboratory Question.

1. Provide calculations for all dilutions to be made in this lab.

Materials

Apparatus
Safety goggles
Lab apron
Buret
Buret clamp
Ring stand
250-mL beaker
500-mL beaker
Filter paper
Collecting bottle
Funnels (2)

Reagents
Rainwater
Soil
Calcium carbonate
NaOH (0.100 *M*)
Phenolphthalein

Safety
1. Safety goggles and a lab apron must be worn at all times in the laboratory.
2. If you come in contact with any solution, wash the contacted area thoroughly.

Procedure

Part 1   Determining the pH of Rainwater

1. Obtain a sample of rainwater by placing a funnel in the mouth of a collecting bottle and placing the bottle and funnel outside while it is raining.

2. Make 50.0 mL of 1.00×10^{-4} M NaOH from the 0.100 M NaOH provided by your teacher.

3. Titrate 50.0 mL of rainwater with 1.00×10^{-4} M NaOH. Use phenolphthalein indicator. Titrations were introduced in Chapter 15. Record the volume of NaOH.

4. If the titration requires less than about 5 mL of the 1.00×10^{-4} M NaOH, dilute the NaOH solution so the titration of 50.00 mL rainwater will require about 20 mL of aqueous NaOH (you will need to calculate the exact concentration of the NaOH).

5. Titrate the new sample with the new NaOH and record the volume of NaOH.

Part 2   Determining the Effect of Soil

1. Obtain a soil sample from home or outside your school. Collect about 200 mL of soil in a 500-mL beaker.

2. Place the soil in a funnel fitted with filter paper.

3. Pour the rainwater through the soil and collect it into a 250-mL beaker

4. Collect the rainwater that has been poured through soil (filter again if there is still soil in the sample).

5. Titrate 50.0 mL of rainwater with NaOH of the same concentration used in part I (either step 3, or step 5, if you needed to perform steps 4 and 5). Use phenolphthalein indicator. Record the volume of NaOH used.

Part 3   Increasing Soil Capacity

1. Crush and powder 5 grams of calcium carbonate. Mix with the same amount of soil as in Part II. (mix until homogeneous)

2. Place the soil in a filter.

3. Pour the rainwater through the soil and collect in a 250-mL beaker.

4. Collect the rainwater that has been poured through the soil (filter again if there is still soil in the sample).

5. Titrate 50.0 mL of this rainwater with NaOH of the same concentration in Part I (either step 3, or step 5 if you need to perform step 5). Use phenolphthalein indicator. Record the volume of NaOH.

Cleaning Up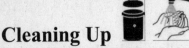

1. Empty the buret and rinse thoroughly. Clamp it back in the buret clamp upside down to drain.
2. Dispose of all chemicals as instructed by your teacher.
3. Wash your hands thoroughly before leaving the laboratory.

Analysis and Conclusions

Complete the **Analysis and Conclusions** section for this experiment either on your Report Sheet or in your lab report as directed by your teacher.

Part 1

1. What is the number of moles of NaOH used in the titration?

2. What is the number of moles of acid in your rainwater sample?

3. Determine the concentration of H^+ in your rainwater sample.

4. Calculate the pH of your rainwater sample

5. Collect class data.

Parts 2 and 3

6. Determine the pH of rainwater after it flows through untreated soil.

7. Determine the pH of rainwater after it flows through soil mixed with calcium carbonate.

8. What is the normal range of pH for rainwater? Use class data.

9. Does the untreated soil neutralize the rainwater?

10. Does soil mixed with calcium carbonate neutralize the rainwater?

11. Which soil (untreated or mixed with calcium carbonate) neutralized the rainwater better?

Something Extra

1. Does boiling affect the pH of rainwater? Boil and cool rainwater and test it. Explain your results.

2. Does potting soil have an effect on the pH of rainwater? Test it and compare your results to the soil you gathered.

Experiment 68

Indicators

Problem

How can indicators be used to gain information regarding the pH of aqueous solutions?

Introduction

The pH scale is used to indicate the concentration of hydrogen ions (acidic solutions) or hydroxide ions (basic media) in aqueous solutions. Indicators are dyes, usually of plant origin, that undergo a color change when the pH of the system changes. Depending on the particular indicator, the color change may occur in the acidic range (0≤pH≤6), near neutral (~pH 7), or in the basic range (8≤pH≤14). Because most of these dyes have very intense colors, typical concentrations of indicator solutions are only 0.5-1.0%, and it is normal to use only one or two drops of these very dilute solutions when testing with them.

Recall that acids are sources of hydrogen ion, $H^+(aq)$; bases, on the other hand, are sources of hydroxide ion, $OH^-(aq)$. Strong acids, such as hydrochloric acid, $HCl(aq)$ are essentially 100% ionized. Since there is a relatively large concentration of hydrogen ions in solutions of strong acids, you might expect them to have very low pH values. Weak acids, such as acetic acid, $HC_2H_3O_2$, are only partially separated into ions, so their pH values typically run 2 to 3 pH units higher. Thus, it should be possible to find an indicator that will make it possible to distinguish between strong and weak acids. Likewise, it should be possible to distinguish between weak and strong bases, in this case, sodium hydroxide and aqueous ammonia. The correct formula for a solution of ammonia in water is $NH_3(aq)$. At first, it would seem that this isn't a base at all, because there is no hydroxide in the formula, but dissolved ammonia molecules react with the water molecules according to the equation

$$NH_3(aq) + H_2O(l) \rightarrow NH_4^+(aq) + OH^-(aq)$$

This explains why bottles of aqueous ammonia are often labeled "ammonium hydroxide."

There will be two parts to your investigation. First, you will test 0.10 M solutions of four acid and base solutions with each of four indicators, in a effort to see which indicators best distinguish between the four categories mentioned above: strong acid, weak acid, strong base, and weak base. You will also carry out a procedure known as a *titration*, in which you start with a fixed amount of strong acid (HCl), to which a drop of one of the indicators has been added. To this mixture, you will add a strong base (NaOH), one drop at a time, until the indicator changes color. In addition, you will also do titrations using a mixture known as *universal indicator*, which is a mixture of other indicators, so changes color more or less regularly as the pH ranges from a low of 1 or 2, all the way up to 13 or 14. A color chart is provided with the universal indicator to allow you to determine the pH represented by any particular color.

Prelaboratory Assignment

✓ Read the **Introduction** and **Procedure** before you begin.
✓ Answer the Prelaboratory Questions.

1. Suppose the instructions told you to empty the contents of the well plate into a waste beaker at the conclusion of Part 1 of the procedure. Do you think the solution in the waste beaker would be acidic, basic, or neutral? Give a reason for your prediction.

2. Describe a way to test your hypotheses from Prelaboratory Question 1, using only the materials provided for this experiment.

Materials

Apparatus

24-well test plate
pH test paper
Beaker, 20 mL (for waste)
Distilled water in wash bottle
Toothpicks for mixing
Safety goggles
Lab apron

Reagents (In microtip pipets)

Acetic Acid, $HC_2H_3O_2$, a weak acid, 0.1 M
Ammonia, NH_3, a weak base, 0.1 M
Hydrochloric Acid, HCl, a strong acid, 0.1 M
Sodium Hydroxide, NaOH, a strong base, 0.1 M
Indicators
    Bromcresol Purple
    Congo Red
    Indigo Carmine
    Phenolphthalein
    Universal Indicator

Safety

1. Acids and bases are corrosive to skin and clothing, even when present in small amounts and in low concentrations. Avoid contact with your skin and wipe up all spills with large amounts of water.
2. Indicators can cause permanent stains in clothing. Many are made with alcoholic solvents, so are quite flammable. Keep them away from open flames.
3. Safety goggles and a lab apron must be worn at all times in the laboratory.

Procedure

If you are not using a Report Sheet for this experiment, design a Data Table with the following headings:

Solution Tested	Congo Red	Indigo Carmine	Bromcresol Purple	Phenol- phthalein	Universal Indicator

Part 1 Indicator colors

1. Arrange your 24-well plate with 4 wells as rows (A,B,C,D) and 6 wells as columns (1,2,3,4,5,6).

A1	A2	A3	A4	A5	A6
B1	B2	B3	B4	B5	B6
C1	C2	C3	C4	C5	C6
D1	D2	D3	D4	D5	D6

2. Place 20 drops of 0.10 M NaOH in each of the four wells of column 1 (A1 – D1) of your 24-well plate. Add one drop of one of the indicators to the well A1, one drop of the second indicator to well B1 and so on. Do not use the universal indicator at this point. To mix the liquids, stir gently with a toothpick. Note the color of the mixture and record it in the Data Table.
Note: Do not empty the wells, as you will need some of the solutions for Part 2.

3. Place 20 drops of 0.10 M NH$_3$ in each of the four wells of column 2 (A2 – D2), and add indicators as you did in step 1. Be sure to add them in the same order. That is, if well A-1 has congo red, A-2 should also.

4. Continue in this fashion, placing 0.10 M HC$_2$H$_3$O$_2$ in wells A-3 through D-3, and 0.10 M HCl in A-4 through D-4 and then adding the appropriate indicators.

If you are not using a Report sheet for this experiment design a Data Table with the following headings for Part 2:

Indicator Tested	Drops NaOH needed	Color of Univ. Ind.	Approx. pH

Part 2   Indicator pH Ranges

1. To well A-4, containing hydrochloric acid and one of the indicators, add sodium hydroxide, one drop at a time, until the indicator changes color. Note and record the number of drops needed to cause the color change. Use a toothpick for stirring.

   Repeat the process for well B-4, C-4, and D-4, so that the number of drops needed to change the color of each indicator has been recorded. It is possible that more than 20 drops of NaOH will be needed for one (or more) of the indicators.

2. Place 20 drops of HCl in each of the 4 wells of column 5 of your well plate (A5 – D5). Add 1 or 2 drops of Universal Indicator to each of these wells. Now to well A5, add the same number of drops of NaOH needed to change the color of the indicator in well A4. Record the color of the Universal Indicator. Now do the same process for B5 using the same number of drops as required for B4, and so on for wells C5 and D5.

Cleaning Up

1. Clean the well plate by rinsing the contents down the drain with a large volume of water. Rinse the clean well plate with distilled water, then leave it on paper towels or a rack to dry.
2. Wash your hands thoroughly before leaving the laboratory.

Analysis and Conclusions

Complete the **Analysis and Conclusions** section for this experiment either on your Report Sheet or in your lab report as directed by your teacher.

1. Answer the following questions based on your data and on your knowledge of acids and bases. Some questions may have more than one correct answer. If more than one indicator will accomplish a particular task, list all those that will. In each case, justify your choices by citing evidence from the data table.

   a. Which of the four indicators could be used to tell a strong acid from a strong base?

   b. Which of the four indicators could be used to tell a weak acid from a weak base?

   c. Which of the indicators (if any) could be used to tell a strong acid from a weak acid?

   d. Which of the indicators (if any) could be used to tell a strong base from a weak base?

2. Use the color chart provided for the Universal Indicator to estimate the approximate initial pH of each of the acid and base solutions you used.

3. Use your information from Part 2 to determine the approximate pH range in which each of the four indicators from Part 2, Step 2 changes color.

4. Use your results from Part 2 to explain your conclusions from Part 1.

Experiment 69

Analysis of Vinegar

Problem
How can the acid content of vinegar be determined experimentally?

Introduction
Ordinary "white" vinegar is an aqueous (water) solution of acetic acid which often carries the notation that the acidity has been reduced to 5% with water. Flavored products, like apple cider vinegar, red wine vinegar and balsamic vinegar have other ingredients and flavorings, but even they are essentially acetic acid in water.

In order to ensure that the acidity is at the desired level, periodic routine analyses are run. A common method for such analyses is a *titration*, in which a strong base of known concentration is used to determine the concentration of the acid by allowing the solutes of the two solutions to react with each other. In a titration, a solution (called the *titrant*) is added at a controlled rate to a known amount of the solution to be analyzed. Addition continues until the reaction is complete.

An *indicator* is often used to determine when all of the solute of the solution being tested has reacted. Indicators signal that a reaction is complete by changing color. *Phenolphthalein*, the indicator that you will use in this experiment, is colorless in acidic or neutral solutions, but turns bright magenta with the slightest excess of base. The first drop of base that causes the color to persist signals the end of the titration. The equation for the reaction between the sodium hydroxide and the acetic acid in the vinegar is:

$$HC_2H_3O_2\,(aq) + NaOH\,(aq) \quad \rightarrow \quad Na\,C_2H_3O_2\,(aq) + H_2O\,(l)$$

The hydroxide ion of the base (NaOH) reacts with a hydrogen ion from the acid ($HC_2H_3O_2$) to form water. These reactions are called neutralization reactions, because the acid and the base neutralize each other, producing water. Notice that only one hydrogen atom on the acetic acid molecule reacts with OH$^-$. Many common organic acids contain some hydrogen atoms that are *acidic* and others that are not. The difference need not concern you here; it is enough to know that acidic hydrogens will react with a base, but the others do not.

You will prepare a sodium hydroxide solution of known concentration, then use that solution to analyze the acid content of white vinegar. Once you determine the molar concentration (molarity) of the sodium hydroxide, you will convert the concentration to units of moles of NaOH per gram of solution.

Because in the experiment both the mass and the concentration of the sodium hydroxide titrant are known, the number of moles of NaOH that reacts can be calculated. As the equation shows, acetic acid and sodium hydroxide react in a 1:1 mol ratio, so you can also determine the number of moles of acetic acid present in the sample which can then be converted to mass in grams.

You will carry out four trials for the analysis. The amounts of the two solutions used in each of the titrations will be determined by weighing the pipettes before and after each of the titrations.

Prelaboratory Assignment

✓ Read the **Introduction** and **Procedure** before you begin..
✓ Answer the Prelaboratory Questions.

1. Write the equation for the reaction between acetic acid and NaOH. Use the structural formula for acetic acid found in Chapter 16. In the formulas of the reactants, circle the atoms that form water.

2. What is the purpose of the indicator? How does it tell you when a titration is complete?

3. Give two reasons for *Safety Special Note #2*. (**Hint:** Consider both your own experiment and that of the person who uses the balance after you.)

4. What does the parenthetical expression (± 0.001 g) mean in Step 1 of Part 1 of the Procedure?

5. Read Step 4 of Part 1 of the **Procedure**. Answer the question that appears in parentheses.

Materials

Apparatus
Milligram balance
Beaker, 30- or 50-mL
Graduated cylinder, 25-mL
4 Erlenmeyer flasks, 10-mL (or small beakers)
2 microtip pipets, labeled NaOH and Vinegar
Distilled water wash bottle
Safety goggles
Lab apron

Reagents
Sodium hydroxide, solid pellets
Phenolphthalein indicator,
  in microtip pipet
15 mL vinegar (includes some for Cleaning Up)

Safety

1. Laboratory goggles and a lab apron must be worn at all times in the laboratory.
2. Sodium hydroxide is highly caustic (see Special Notes below). Avoid contact with skin and clothing and wipe up all spills.

Special Notes:

1. **Working With Solid NaOH:** The pellets are highly caustic. They will harm skin and clothing, and they will attack metal surfaces such as the pan of the balance. In addition, they rapidly absorb moisture from the air, so it is critical that Part A of the procedure be done quickly and efficiently.

2. **Weighing Glassware:** Always dry the outside surfaces of beakers, flasks, and other pieces of glassware before you place them on the balance pan.

3. **Weighing Solutions in Pipets:** The pipet cannot be weighed accurately if it is in contact with any surface of the balance other than the balance pan. For this reason, and to protect against spills and leaks, the pipet is placed **tip up** in a small paper or plastic cup which sits on the pan of the balance.

Procedure

Part 1:   Preparation of the sodium hydroxide solution.

1. Determine the mass of a small beaker (±0.001 g). Add 3-4 sodium hydroxide pellets, quickly close the NaOH container, then reweigh the beaker and pellets. Enter both masses in a Data Table. The mass of NaOH should be on the order of 0.3-0.5g.

2. Add 5-6 mL of distilled water to the beaker, then swirl the beaker gently to dissolve the pellets. Carefully feel the bottom of the beaker; is the dissolving exothermic or endothermic?

3. Determine the mass of your clean, dry 25-mL graduated cylinder. Record the mass in the Data Table.

4. When the pellets of sodium hydroxide have completely dissolved, pour the solution from the beaker into the graduated cylinder. Use 4-5 mL of distilled water from your wash bottle to rinse the beaker, then transfer the rinsings to the contents of the graduated cylinder. Rinse the beaker again with 4-5 mL of distilled water, and again transfer the rinsings to the graduate. Continue the rinsing and transferring until the total volume of solution is *exactly* 25.00 mL. **Do not go above the 25.00 mL** line on the graduate. (Why not?)

5. Determine the total mass of the graduated cylinder and the solution. Be careful to dry the outside and the base of the graduate before you place it on the balance pan.

Part 2: Titration of Vinegar.

1. Rinse *and dry* your small beaker, then return the standard sodium hydroxide solution to the beaker. Label a clean microtip pipet, "NaOH," then fill it from the solution in the beaker. Weigh the pipet and record the mass in the Data Table.

2. Fill a second microtip pipet, labeled "Vinegar," or simply, "VIN," from the commercial vinegar bottle. Weigh the filled pipet and record its mass.

3. Weigh a clean 10-mL Erlenmeyer flask. Transfer some of the vinegar solution to the flask, then reweigh the pipet. The mass of vinegar used should be between 0.8 and 1.2 grams; if it is less than 0.8 g, add a bit more, then reweigh the pipet with the remaining vinegar. Record the mass of the pipet and contents in the Data Table.

4. Add 1 drop of phenolphthalein indicator from the phenolphthalein pipet (labeled "PHTH") to the contents of the flask and swirl gently.

5. Dropwise and with swirling, add your NaOH solution a few drops at a time until you get a magenta color that does not fade with mixing and that lasts at least 20 seconds. The lighter the pink color, the better. Reweigh the NaOH pipet and record the mass in the Data Table.

6. Repeat the titration three more times, recording the masses of the pipets before and after each trial. You will probably need to refill one or both pipets from time to time. Try to plan so that you don't have to refill in the middle of a trial.
   Note: Use a clean flask for each titration; if necessary, wash flasks between trials, then rinse with distilled water. You need not dry the flasks.

Cleaning Up

The contents of your titration vessels are safe to be rinsed down the drain with water. However, any unused sodium hydroxide solution remaining in your pipet should be neutralized.

1. Empty the pipet into the beaker or flask containing any unused sodium hydroxide remaining from the experiment and one drop of phenolphthalein.

2. Add vinegar until the pink color of the phenolphthalein *disappears*; the color change signals that neutralization is complete. Dispose of this neutral solution down the drain with plenty of water.

3. Rinse all equipment thoroughly with water, followed by distilled water, then leave it on paper towels to drain.
4. Wash your hands before leaving the laboratory.

Analysis and Conclusions

Complete the **Analysis and Conclusions** section for this experiment either on your Report Sheet or in your lab report as directed by your teacher. Show samples of all calculations.

1. Calculate the mass of sodium hydroxide used in preparing your solution. Convert this to moles.

2. Determine the molarity of your sodium hydroxide solution.

3. Determine the mass of the sodium hydroxide solution you prepared, then use that mass to determine the density of your sodium hydroxide solution.

4. Calculate the concentration of the NaOH solution in moles of NaOH per gram of solution.

5. From your data, calculate the mass of NaOH solution used, the number of moles of NaOH used, and the number of moles of acetic acid that must have been present for each of your four titrations. Show your work for trial 1; enter the results for all four titrations in a Data Table.

6. Determine the mass of acetic acid present in each of your four titration samples. Find the percent of acetic acid in each vinegar sample, by dividing the mass of acetic acid present by the mass of vinegar used. Convert the decimal fraction to a percent. Present your results in a Summary Table.

7. Calculate an average value for the mass percent of acetic acid in the vinegar you analyzed. Base your average on the three trials that show the closest agreement; omit the trial that deviates most greatly from the others.

8. Calculate the deviation from the average for each of the four trials, then calculate the average deviation for the three trials that show the best agreement.

9. Report the mass percent of acetic acid in the vinegar as a percent ± average deviation. Assuming the density of white vinegar to be 1.0 g/mL, calculate:
   a. the mass of 1.0 L of white vinegar
   b. the mass of acetic acid in 1.0 L of vinegar
   c. the number of moles of acetic acid in 1.0 L of vinegar
   d. the molar concentration (molarity) of acetic acid in white vinegar

Something Extra

Sodium hydroxide reacts with the moisture and carbon dioxide in the air, so it is very likely that the mass you reported in question 1 of Analysis and Conclusions is not all sodium hydroxide. Find out how to use potassium acid phthalate, KHP, to standardize a sodium hydroxide solution. Then, with your teacher's permission, repeat the experiment, this time standardizing the solution before proceeding to Part 2.

Experiment 70

Quantitative Titration

Problem

How can the amount of acid in an unknown solution be determined? What is the percentage of acid in vinegar or lemon juice?

Introduction

Titration is the name given to the process for determining the volume of a solution needed to react with a given mass or volume of some particular chemical sample. This procedure has many applications, but is especially useful in the reaction between an acid and a base. In this investigation you will use a titration to determine the strength of an unknown acid. A prepared solution of a base will be systematically added to the acid solution. As the OH^- ion in the base reacts with the H^+ ions in the acid solution, water forms. The titration is complete when the number of OH^- ions added is exactly equal to the number of H^+ ions that are present in the acid. According to the formula, $M_{acid}V_{acid} = $ moles of $H^+ = $ moles of $OH^- = M_{base}V_{base}$, the molarity of the unknown acid can be determined. The end point of the titration will be signaled by the color change of an indicator. The indicator used in this titration is called phenolphthalein. Phenolphthalein exhibits a pink color in the presence of a base, but is colorless in the presence of an acid.

Prelaboratory Assignment

✓ Read the **Introduction** and **Procedure** before you begin.
✓ Answer the Prelaboratory Questions.

1. Determine the number of grams of NaOH needed to make 100.0 mL of a 1.00 M NaOH solution.

2. If 24.3 mL of 0.085 M NaOH solution are needed to completely neutralize 15.5 mL of an unknown acid, what is the concentration of the acid solution?

3. What is the $[H^+]$ in each of the following acid solutions?
   a. 0.004 M HNO_3?
   b. 1.33 M $HClO_4$?
   c. 12 M HI?

Materials

Apparatus
Centigram or milligram balance
125-mL Erlenmeyer flasks (2)
100-mL graduated cylinder
Weighing paper
Buret
Ring stand assembly with buret clamp
Safety goggles
Lab apron

Reagents
Solid sodium hydroxide, NaOH
1% phenolphthalein indicator solution
Vinegar, lemon juice, or some other unknown
   acid solution
Distilled water

Safety

1. Wear safety goggles and a lab apron at all times in the laboratory.
2. No food or drink is allowed in the laboratory at any time.
3. Do not handle sodium hydroxide pellets with your fingers; the NaOH is caustic

Procedure

Part 1 - Preparation of the Sodium Hydroxide Solution

1. On a piece of paper, weigh out the number of grams of sodium hydroxide pellets that you have determined are needed to make 100.0 mL of a 1.00 *M* NaOH solution.
   Note: Do not touch the NaOH pellets with your hands; NaOH is very caustic.

2. Place the sodium hydroxide into the 150-mL Erlenmeyer flask and dissolve it in about 50 mL of distilled water.

3. When the pellets are fully dissolved, place the solution into a clean, 100-mL graduated cylinder. Carefully add distilled water until the volume is exactly 100 mL.

4. Obtain a buret and rinse it with distilled water. Be careful to fill the buret at least half-way, swirl the water along the sides, and allow some of the water to run through the stopcock.

5. Now rinse the buret with a few milliliters of the NaOH that you have prepared. Allow a little bit of NaOH to run through the stopcock. Empty the buret of extra NaOH.

6. Clamp the buret to the ring stand.

7. Fill the buret with the NaOH solution and record the volume of solution in the buret. You need not fill it completely to the zero mark. **Note:** Be sure to read the buret at eye-level and to read the bottom of the meniscus of the liquid.

Part 2 - Titration of the Unknown Acid with the NaOH Solution

1. Obtain a sample of unknown acid as directed by your teacher. Measure approximately 20 mL of the acid solution into a clean, graduated cylinder. Record the exact volume, then transfer the acid to a clean 125-mL Erlenmeyer flask.

2. To the acid solution add six to eight drops of phenolphthalein indicator. Swirl to mix.
   Note: The solution should be colorless at this point.

3. Place the flask containing the acid under the tip of the buret. It may be helpful in detecting the end point to place the flask on a piece of white paper.

4. Open the stopcock slightly and allow the NaOH to flow slowly into the acid. As the NaOH drips into the acid, swirl the flask to completely mix the acid and base together. Continue dripping the NaOH into the flask until the first persistent sign of pink appears.
   Note: A bright magenta color is not satisfactory! You have gone past the endpoint and must repeat the trial.

5. Note and record the final volume of the base in the buret.

6. Repeat the titration with another 20.0 mL of the same acid. Repeat the titration a third time. Each time, be careful to record the initial volume and the final volume of the NaOH in the buret.

Cleaning Up

The contents of your titration vessels are safe to be rinsed down the drain with water. However, any unused sodium hydroxide solution remaining should be neutralized.

1. Empty the NaOH into a beaker or flask, add any unused sodium hydroxide remaining from the experiment and one drop of phenolphthalein.
2. Add vinegar until the pink color of the phenolphthalein *disappears*; the color change signals that neutralization is complete. Dispose of this neutral solution down the drain with plenty of water.
3. Rinse your buret well with water, and leave it clamped upside down to drain.
4. Return all glassware to its proper location.
5. Wash your hands before leaving the laboratory.

Analysis and Conclusions

Complete the **Analysis and Conclusions** section for this experiment either on your Report Sheet or in your lab report as directed by your teacher.

1. For each of the three titrations, determine the number of milliliters of NaOH required to reach the end point.

2. Calculate the molarity of the unknown acid in each of the three samples.

3. Determine the average molarity of the unknown acid. Report your average answer to three significant figures.

4. What effect would each of the following have on the molarity of your acid solution?
   a. Rather than the desired light pink color, a bright magenta color marked the endpoint of the titration.
   b. Ten drops of phenolphthalein, rather than 6-8, were used in the titration.
   c. The flask was not swirled during the titration, and the experiment was stopped at the first sign of pink.

Something Extra

1. Using chemical resources, investigate other common indicators. Do all indicators change color at the same pH?

2. Repeat this experiment to compare the amount of acid in lemon juice, orange juice, or grapefruit juice samples (or in various types of vinegar). Report your findings.

3. Using chemical resources, identify the primary acid found in each of the following:
   a. Aspirin
   b. Vinegar
   c. Milk

Experiment 71

Conductivity Titrations
A CBL Investigation

Problem

How can the progress of a precipitation titration be followed?

Introduction

Acid-base reactions are commonly studied by titrations in which the progress of the reaction is monitored by using a pH meter. On the other hand, if the goal is simply to determine when the reaction is complete, an indicator can be used. Recall that indicators are compounds that change color when the pH of the system reaches a certain value. In precipitation reactions, it is the formation of an insoluble solid that is the driving force; the pH may not change at all.

In precipitation reactions what does change is the total number of ions in the system. The more ions there are, the better the solution is able to conduct an electric current. So one way to monitor such a reaction is by observing changes in the *conductivity* of the system.

The reaction you will study in this experiment is the one that occurs when a dilute oxalic acid solution is added to a solution of lead(II) acetate. Like many compounds containing the lead(II) ion, lead(II) oxalate has a low solubility. Also, acetate is a stronger base than the oxalate ion. Thus, mixing solutions of lead(II) acetate and oxalic acid results in the formation of a white precipitate of lead(II) oxalate, PbC_2O_4, along with the weak acid, $HC_2H_3O_2$. Because you are going from strong electrolyte to weak acid and insoluble ionic salt, the conductivity drops steadily, reaching a minimum at the equivalence point. Beyond the equivalence point the conductivity increases again.

Your teacher will provide you with solutions of oxalic acid and lead(II) acetate. You will measure the initial conductivity of the solution, then add small amounts of oxalic acid, $H_2C_2O_4$, measuring the conductance after each addition. The molecular equation for the reaction that takes place is:

$$Pb(C_2H_3O_2)_2(aq) + H_2C_2O_4\ (aq) \rightarrow PbC_2O_4\ (s) + 2\ H\ C_2H_3O_2\ (aq)$$

The Vernier Conductivity Probe is designed to measure the electrical conductivity of a solution, and so can be used to see how the conductivity changes as one reactant is added to another. In this system, the lead(II) oxalate precipitates from the solution, and the other product (acetic acid) is a weak acid and thus has very low conductivity. Therefore, the ability of the solution to conduct an electric current should reach a minimum when the number of moles of lead(II) ion originally present is equal to the number of moles of oxalic acid added. This point is known as the *equivalence point*.

Prelaboratory Assignment

✓ Read the **Introduction** and **Procedure** before you begin.
✓ Answer the Prelaboratory Questions.

1. The equation for the reaction you are investigating is given in molecular form in the Introduction. Present the same reaction as:
   a. a complete ionic equation (assume the oxalic acid dissociates into two hydrogen ions an oxalate ion).
   b. a net-ionic equation.

2. Why is it necessary to filter out the precipitated solid remaining after the completion of the experiment?

3. The lead(II) acetate solution has an approximate concentration of 0.010 M. What volume of 0.10 M oxalic acid should be needed to reach the equivalence point? Show calculations to defend your prediction.

Materials

Apparatus
TI 83+, or similar graphing calculator
CBL, CBL-2, or Lab Pro
CHEM/BIO or DataMate
Vernier Conductivity Probe
Link cable
CBL-DIN adapter (older probes only)
100-mL beakers (2)
Calibrated plastic micropipet or buret
Magnetic stirrer and stirring bar, if available
Ring stand and iron ring
Funnel
Filter paper
Gloves
Apron
Safety goggles

Reagents
Lead(II) acetate, 0.010M(aq)
0.10 M oxalic acid, $H_2C_2O_4$(aq)
Distilled water (wash bottle)

Safety

1. Lead compounds are very toxic. Avoid all contact with the lead(II) acetate solution and with the solution in the beaker after the reaction.
2. Oxalic acid, even in low concentration, is corrosive to skin and clothing.
3. Clean up all spills with large amounts of water.

Procedure

If you are using original CBL, the CHEM/BIO application (APPS) contains the necessary data acquisition program. If you are using CBL2 or LabPro, the DataMate was application can be loaded directly from the interface. If you have not yet installed the appropriate program into your calculator, you must do so before you begin. You can also obtain it from another calculator.

Setting Up the System

1. Connect the calculator to the CBL using the link cable. Connect the conductivity probe to the interface, using a CBL-DIN adapter if necessary. **Figure 1** illustrates the manner in which the apparatus is to be set up. It assumes that you will use a magnetic stirrer and stirring bar, and that you will add the oxalic acid from a calibrated pipet. If you do not have a magnetic stirrer, the beaker contents will have to be stirred by hand, using a clean glass stirring rod. If addition of the acid is to be done by buret, it should be clamped in position next to the beaker. Turn on the calculator and interface, and open DataMate (or other application). The conductivity probe is connected to Channel 1 of the CBL. Set the conductivity probe to 2000 mic.

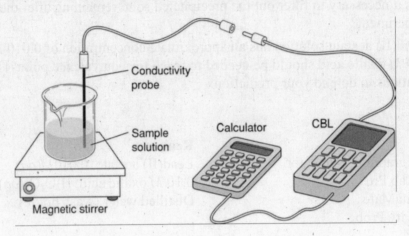

Figure 1
Apparatus for conductivity titration

Conducting the Titration

Using original CBL, open the CHEM/BIO program from APPS. Select SET UP PROBES; you have one probe, and it is to measure conductivity. Choose the TRIGGER/PROMPT option.

For CBL2 or LabPro, DataMate should recognize the conductivity probe; if it does not, use SETUP to show the conductivity probe in Channel 1. The mode is EVENTS WITH ENTRY.

2. Place the stirring bar in the beaker and add 20.0 mL of lead(II) acetate solution. Place the beaker on the stirrer and start the bar spinning; adjust the rate so that the surface of the liquid just barely shows a vortex. Carefully lower the probe so that it is submerged in the solution but is not in danger of being struck by the stirring bar. Push ENTER to get your first reading. When you see PROMPT? (CBL) or ENTER VALUE (CBL2, LabPro) enter 0 (zero). Accept the option to COLLECT MORE DATA.

3. Use the pipet (or buret) to add 0.50 mL of the 0.10 M oxalic acid. Read the conductivity again; this time enter 0.50 when prompted; again, you want to COLLECT MORE DATA.

4. Take additional readings after each addition of 0.50 mL of the acid solution. Continue in this fashion until you have added a total of 5.00 mL, or until the conductivity has risen to the level of the initial (0 mL) reading.

Cleaning Up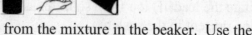

1. Raise the probe from the mixture in the beaker. Use the wash bottle to rinse the probe with distilled water. Allow the rinsings to fall into the beaker.
2. Use metal forceps or tweezers to remove the stirring bar from the beaker. Hold it over the beaker while you rinse it with the wash bottle.
3. Set up for filtration, using an iron ring and ring stand. Place your second beaker under the funnel, then filter the mixture in the beaker where the reaction took place. To do this, swirl the contents to make a slurry of the precipitate, then pour into the filter.
4. When filtration is complete, carefully remove the filter with the lead(II) oxalate precipitate from the funnel and place it in the location designated by your teacher.
5. The filtrate (liquid) in the receiver beaker contains a very dilute solution of acetic acid and can be rinsed down the drain safely, using large amounts of water.
6. Wash both beakers, the stirring bar, and any other apparatus that came in contact with the lead solution or precipitate. Use soap and water.
7. After removing your gloves, wash your hands thoroughly with soap and water.

Data/Observations

1. Using either the TRACE function, or STAT/EDIT from your calculator, summarize your data in a table with the headings:

Volume of $H_2C_2O_4$ (mL)	Conductivity Reading

2. Use graph paper to plot your data of conductivity vs. volume of $H_2C_2O_4$. Assume that the rounded V-shaped curve is actually two intersecting straight lines, one with negative slope and the other with a positive slope. Draw the best-fit straight line for each side. Extend each so that the two intersect. The intersection point is your experimental value for the equivalence point of the reaction. Be sure to label the axes.

Analysis and Conclusions

Complete the **Analysis and Conclusions** section for this experiment either on your Report Sheet or in your lab report as directed by your teacher.

1. Describe the shape of the curve. Include a description of the relative slopes of the two intersecting lines.

2. Identify the volume of $H_2C_2O_4$ that was needed to reach the equivalence point (this may or may not be a whole number, and almost certainly will not coincide with any of your data points).

3. Compare the value from the preceding question with your prediction in **prelaboratory question #3**. Suggest a possible explanation for any significant difference.

4. There are two products to the reaction: lead(II) oxalate, which is insoluble, and acetic acid, which is a weak electrolyte. Account for the fact that once the equivalence point is reached, the conductivity increases again as $H_2C_2O_4$ is added.

Something Extra

1. As you add the oxalic acid and precipitate the lead(II) ion from solution as $PbC_2O_4(s)$ you are not changing the total amount of lead(II) ion present in the system.

   a. Use the volume and molarity of the oxalic acid to determine the number of moles of oxalate ion added at the equivalence point.

   b. The result from **a.** must also be the number of moles of lead(II) ion that were present. Why?

   c. Use the number of moles of lead(II) ion that were present in the 20.0 mL of the lead(II) acetate solution to calculate the actual molarity of the original lead(II) acetate solution.

Experiment 72

Acids, Bases, and Buffers

Problem

How does the presence of a buffer affect the progress of a pH titration?

Introduction

Acetic acid is a weak acid; that is, only a very small percentage of the acetic acid molecules are present in ionic form at any one time. At any given instant, some of the molecules are separating into ions, but at the same time, previously-separated ions are reuniting, forming acetic acid molecules once again. This results in a situation, in which molecules are separating into ions at exactly the same rate that ions are rejoining to make molecules. Such a balance of opposing processes is described as being a *reversible reaction*, and is represented by a double arrow, thus:

$$HC_2H_3O_2(aq) \rightleftarrows H^+(aq) + C_2H_3O_2^-(aq) \qquad (1)$$

You will learn more about such reversible systems in Chapter 17; for now, all you need to know is that this is what is happening in an acetic acid solution such as vinegar, and that there are far more undissociated molecules than there are ions at any given time as indicated by the different sizes of the arrows.

As you learn in Chapter 16, the pH scale serves as a measure of the concentration of hydrogen ions in an aqueous solution. We represent this concentration by $[H^+]$; the square brackets are understood to refer to the concentration (molarity) of the ion or molecule they enclose. Because acetic acid does not have many free hydrogen ions present at any one time, the pH is not as low as it would be for a solution of a strong acid, such as hydrochloric acid, $HCl(aq)$, which is 100% broken down into ions.

Because the acetate ion, $C_2H_3O_2^-$, has a strong tendency to combine with hydrogen ions, as shown by equation (1) above, acetate acts as a base. That is, it acts as a proton acceptor. Sodium acetate, $NaC_2H_3O_2$, is a strong electrolyte. It is totally dissociated into ions when dissolved in water.

$$NaC_2H_3O_2(aq) \longrightarrow Na^+(aq) + C_2H_3O_2^-(aq) \quad (2)$$

An interesting type of mixture, which you will investigate in this experiment, is called a *buffer*. Buffers are of two types. An acidic buffer consists of a weak acid, such as acetic acid, mixed with a salt that contains the anion of the acid (for acetic acid, the acetate ion, $C_2H_3O_2^-$). A basic buffer consists of a weak base, such as ammonia, $NH_3(aq)$, and a compound that contains the cationic form of the base (in this case, NH_4^+).

You will investigate an acidic buffer, consisting of acetic acid and sodium acetate, to see how the buffered system responds when aqueous sodium hydroxide is added. You will compare this result with the behavior you observe when the sodium hydroxide is added to a solution that contains only acetic acid.

Prelaboratory Assignment
  ✓ Read the **Introduction** and **Procedure** before you begin.
  ✓ Answer the Prelaboratory Questions.

1. Calculate the mass of sodium acetate you would need to prepare 30.0 mL of 0.10 M $NaC_2H_3O_2$ solution. Show your calculations.

2. Acids are proton donors; bases are proton acceptors. In these definitions, the "protons" are actually hydrogen ions. Why are hydrogen ions referred to as protons?

3. Ammonia is a weak base. Write the net ionic equation for the reaction in which an ammonia molecule accepts a proton (hydrogen ion) from water.

Materials

Apparatus
Buret, 50-mL
pH meter, or CBL2 or LabPro
  with pH probe and pH amplifier
Beaker, 150-mL (2)
Magnetic stirrer and stirring bar (optional)[1]
250-mL (or larger) beaker, for rinsing
Safety goggles
Lab apron

Reagents
Acetic acid, $HC_2H_3O_2$, 0.10 M
Sodium hydroxide, NaOH, 0.10 M
Sodium acetate, $NaC_2H_3O_2$, solid
Sodium acetate, $NaC_2H_3O_2$, 0.10 M
Distilled water (wash bottle)
Hydrochloric acid, HCl, 0.10 M

Safety

1. Sodium hydroxide is corrosive to skin and clothing. Clean up spills with large amounts of water.
2. Acids are corrosive to the skin and clothing. Be sure to clean up all spills with large amounts of water.
3. Wear laboratory safety goggles and a lab apron at all times in the laboratory.

Procedure
Set up the pH meter as directed by your teacher, or connect the CBL to the pH amplifier and the pH probe. If you are using a CBL unit, connect it to your graphing calculator and launch the DataMate application (APPS).

Note: If you are using CBL2, or LabPro you will not need to record the pH readings directly; the unit will store the data until you need it. You can get the data from List 1 and List 2 (L1 and L2), by selecting STAT/EDIT. Use the EVENTS WITH ENTRY option as the data collection mode.

Part 1   Titration of acetic acid by sodium hydroxide
1. Rinse the buret with a few milliliters of 0.10 M NaOH; be sure to rinse the barrel and the tip of the buret by allowing some NaOH to flow through the buret. Discard the rinsings in your waste beaker, then fill the buret with fresh NaOH.

2. Place 30.0 mL of 0.10 M $HC_2H_3O_2$ in the beaker. Place the beaker on the magnetic stirrer and begin the stirrer. Carefully lower the pH probe into the solution, taking care to position it so the it is not struck by the stirring bar. Read and record the pH of the acetic acid.

---
[1] If you do not have a magnetic stirrer, you will need to swirl the beaker after each addition of NaOH.

3. Add 10.0 mL of 0.10 M NaOH from the buret, allow the solutions to mix thoroughly, then read and record the pH.

4. Continue adding 5.0 mL portions of the NaOH, reading the pH after each addition, until a total of 25.0 mL has been added. Then add the base 1.0 mL at a time, reading the pH after each addition, until the total amount of base added reaches 35.0 mL.

5. Follow the directions for Part 1 in Cleaning Up.
   Note: If you are using CBL, CBL2 or LabPro for data collection, you will need to stop and record your data before proceeding, since it will be overwritten when you do the next titration.

Part 2   Preparing and Titrating the Buffer

1. Weigh out the mass of solid sodium acetate you calculated in Prelaboratory Question 1. Place the solid in one of your two 150-mL beakers. Add 30.0 mL of distilled water and swirl or mix until all of the solid has dissolved.

   In your second beaker, place the same mass of solid sodium acetate but this time in 30.0 mL of the 0.10 M acetic acid. As before, swirl or mix until all of the solid has dissolved. This is the buffered system that you will compare to the unbuffered acid.

2. Use the pH probe to determine the pH of the solution of sodium acetate in distilled water. Record the value. Rinse the pH probe with distilled water, catching the rinsings in your waste beaker.

3. Place the beaker containing the acetic acid/sodium acetate mixture on the magnetic stirrer and place the pH probe in the solution, taking care that the stirring bar cannot strike and damage the probe.

4. Refill the buret with 0.10 M NaOH and titrate the buffer in the same way that you did the acetic acid in Part 1, steps 3-5.

Part 3   Effect of Acid on the buffer.

1. Clean your two titration beakers as you did at the end of Part 1. Fill one with 30.0 mL of distilled water and the other with 30.0 mL of buffer mixture, prepared in the same way as you did for Part 2. Read and record the pH.

2. To one of the beakers from step 1, add 1.0 mL of 0.10 M HCl; read and record the pH. Repeat four more times, for a total of 5.0 mL of the strong acid.

3. Repeat step 2 using the second beaker from step 1. As before, read the initial pH, then add 1.0 mL portions of 0.10 M HCl, up to a total of 5.0 mL of the acid, reading and recording the pH after each addition.

Cleaning Up

Part 1

1. Raise the pH probe from the titration beaker and replace that beaker with your waste beaker. Use the wash bottle to rinse the probe, catching the rinsings in the waste beaker.
2. The contents of the titration beaker are mildly basic, but is safe to rinse down the drain with large amounts of water. Remove the stirring bar, then rinse the beaker well, first with tap water, then with distilled water. It is now ready for use in Part 2.
3. Rinse the stirring bar with distilled water.

Part 2

1. Follow the instructions for Part 1, rinsing all solutions down the drain. Clean the beakers with tap water and distilled water and return them to their proper location.
2. Rinse the stirring bar with distilled water, dry it and return it to where it belongs.
3. Drain the buret into your waste beaker. Add a drop of phenolphthalein to the contents of the beaker. If the indicator turns pink, as it probably will, add acetic acid, a few milliliters at a time, until the pink just disappears. The contents of the beaker may now be rinsed down the drain.
4. Rinse the buret several times with tap water, then with distilled water, being careful to rinse both the barrel and the tip of the buret. Clamp the buret in the inverted position over paper towels so that it can dry.

Part 3

1. If the pH of the solution in your beaker is between 5 and 7, it can be rinsed down the drain with excess water. If it is below 5, add 0.10 M NaOH until the system is in the desired range. This should take very little of the base, so add the NaOH only in small quantities.
2. Clean and put away any remaining glassware. Return the pH meter or CBL materials to the location designated by your teacher.
3. Wash your hands before leaving the laboratory.

Analysis and Conclusions

Complete the **Analysis and Conclusions** section for this experiment either on your Report Sheet or in your lab report as directed by your teacher.

1. Plot your data for each of the two titrations (Parts 1 and 2) on graph paper.

2. Describe the difference between the shapes of the graphs for the buffered and unbuffered systems with sodium hydroxide.

3. Account for the effect of small quantities of a strong acid on the buffered system. Why does the pH not change as rapidly for the buffer as it did for distilled water? What was consuming the acid?

4. Write equations for:
   a. the reaction that occurred in the titrations in Part 1 and 2. (**Hint:** it is the same reaction in both cases.)
   b. the reaction that occurred when HCl was added to the buffer in Part 3.

Something Extra

Write an equation that accounts for the pH of the sodium acetate solution.

Experiment 73

Iodine Clock Reaction

Problem
Do temperature and concentration changes alter the rate of a chemical reaction? If so, to what extent?

Introduction
In this experiment you will investigate the role of concentration and temperature on the rate of a chemical reaction. You will perform various experiments using an interesting reaction called a "clock reaction". You will appreciate the significance of that name after you have completed your first determination.

In order to determine the role of each factor independently, you will vary the concentration of one of the reactants in **Part 1** of the experiment, and you will vary the temperature in **Part 2**. In each case, you will keep other possible variables constant. The clock reaction is performed by mixing the two solutions described below:

1. Solution **A** is a dilute solution of potassium iodate, KIO_3. This is the source of the iodate ion, IO_3^-.
2. Solution **B** contains starch and the other reacting species, the hydrogen sulfite ion, HSO_3^-.

When the two solutions are thoroughly mixed, the ions of the resulting solution undergo the following reactions:

a. $IO_3^-(aq) + 3\,HSO_3^-(aq) \rightarrow I^-(aq) + 3\,SO_4^{2-}(aq) + 3\,H^+(aq)$

b. $5\,I^-(aq) + 6\,H^+(aq) + IO_3^-(aq) \rightarrow 3\,I_2(aq) + 3\,H_2O(l)$

c. $I_2(aq) + HSO_3^-(aq) + H_2O(l) \rightarrow 2\,I^-(aq) + SO_4^{2-}(aq) + 3\,H^+(aq)$

d. $I_2(aq) + $ starch $\rightarrow$ a blue solution!

The reactions take place in sequence. Only when the hydrogen sulfite ion has been entirely consumed in reactions **a**, **b**, and **c** will iodine molecules accumulate in solution and react with the starch in reaction **d**. The appearance of the blue color indicates that all of the hydrogen sulfite has been consumed and I_2 formed remains in solution, available to react with the starch.

Prelaboratory Assignment
✓ Read the **Introduction** and **Procedure** before you begin.
✓ Answer the Prelaboratory Questions.

1. On a molecular level, why should more concentrated solutions react more quickly than less concentrated ones?

2. On a molecular level, why should warm solutions react more quickly than cooler ones?

3. **a.** You are to prepare a solution by mixing 10.0 mL of 0.02 M KIO_3 with 10.0 mL of water. What is the final concentration of the KIO_3 in your diluted solution?

   b. Suppose that you are preparing another solution by mixing 25.0 mL of 0.033 M KIO_3 with 18.0 mL of 0.50 M H_2SO_3. What is the final concentration of the KIO_3 in this solution?

Materials

Apparatus
10-mL graduated cylinder
2 large test tubes (18x150 mm)
Test tube rack
Stop watch or room clock
Wash bottle
2 thermometers
Hot plate (or other heating assembly)
2 400-mL beakers (for water baths)
Safety goggles
Lab apron

Reagents
Solution A (contains iodate ion, IO_3^-)
Solution B (contains bisulfite ion, HSO_3^-)
Distilled water
Ice

Safety
1. Wear safety goggles and a lab apron at all times in the laboratory.
2. Be careful when working with solutions on the hotplate.

Procedure
Part 1   The Effect of Concentration Changes
1. Use a clean graduated cylinder to measure 10.0 mL of solution A.   Pour the solution into a clean test tube. Rinse the graduated cylinder.

2. In a similar manner place 10.0 mL of solution B into another test tube. **Note:** Since you will be using the same volume of solution B in each reaction in this part of the experiment, you may wish to use a small piece of masking tape or a grease pencil to mark the 10.0 mL fill-line on the test tube. By doing so, you will not have to continue to measure the solution B with the graduated cylinder.

3. Make certain that the solutions are at room temperature.

4. At this point, decide which laboratory partner will be the timer and which will be the mixer. Using a stopwatch or a classroom clock with a second hand, the timer should start the watch just as the mixer pours solution A into solution B. As the solutions are poured together, the mixer should pour them back and forth quickly three times to obtain uniform mixing. Time should be recorded from the instant both solutions are in contact. Watch the solution and record the time at the first sign of a reaction.

5. Repeat the experiment to confirm your results.

6. Measure 9.0 mL of solution A into the clean graduated cylinder. Add distilled water from the wash bottle until the volume in the cylinder is exactly 10.0 mL. Pour this solution into a clean test tube.

7. Place 10.0 mL of solution B into another test tube.

8. Mix and time once again as you did in Step 4. Record the time for this reaction.

9. Now into the clean graduated cylinder measure 8.0 mL of solution A. Add distilled water from the wash bottle until the volume in the cylinder is exactly 10.0 mL. Pour this solution into a clean test tube.

10. Place 10.0 mL of solution B into another test tube.

11. Mix and time once again as you did in Step 4. Record the time for this reaction.

12. Continue diluting the solution A as indicated in the following table. Repeat the timing procedure, adding the diluted solution of A to 10.0 mL of solution B. Both solutions should always be at room temperature.

mL Solution A	mL Solution B	mL Distilled Water
7.0	10.0	3.0
6.0	10.0	4.0
5.0	10.0	5.0
4.0	10.0	6.0
3.0	10.0	7.0
2.0	10.0	8.0
1.0	10.0	9.0

13. If this is the last step of the procedure for this class period, clean all test tubes and put all glassware back in their proper locations.

Part 2   The Effect of Temperature Changes
1. Prepare an ice bath by placing a mixture of ice and tap water in a 400-mL beaker.

2. Prepare a hot water bath by warming about 250 mL of tap water in a 400-mL beaker to about 60°C on a hot plate.

3. Use a graduated cylinder to measure 10.0 mL of solution A. Pour this solution into a clean test tube.

4. Use a clean graduated cylinder to measure 10.0 mL of solution B. Pour this solution into a second test tube. **Note:** In this part of the experiment, you will always use 10.0 mL of A and 10.0 mL of B. You may wish to use a small piece of masking tape or a grease pencil to mark the fill-lines on the two test tubes. By doing so, you will not have to continue measuring the volumes of solution with the graduated cylinder.

5. These solutions must be at room temperature before they are mixed. Use the thermometers to record the temperature of each solution. **Note:** If the solutions are not already at the same temperature, you may put both test tubes into a 400-mL beaker about two-thirds full of water at room temperature. Let them stand for several minutes before proceeding.

6. Assign one laboratory partner to be the timer and one to be the mixer. Using a stopwatch or a classroom clock with a second hand, the timer should start the watch just as the mixer pours solution A into solution B. As the solutions are poured together, the mixer should pour them back and forth quickly three times to obtain uniform mixing. Time should be recorded from the instant both solutions are in contact. Watch the solution and record the time at the first sign of a reaction.

7. Repeat this control experiment at least one time.

8. Repeat the experiment, using 10.0 mL A and 10.0 mL B for each of the following temperature ranges.

    Temperatures
    0-5°C
    10-15°C
    20-25°C
    30-35°C
    40-45°C
    50-55°C

In each case, put the test tubes filled with the 10.0 mL A and the 10.0 mL B in either the ice bath or the hot water bath. Leave the test tubes in the water bath for several minutes until their temperatures fall into the appropriate temperature range.

9. Once again, start the stopwatch as solution A is poured into solution B. The mixer should immediately mix the solutions by pouring them back and forth quickly three times. Place the test tube back in the water bath and observe it carefully.

10. Record the time at the first sign of a reaction.

Cleaning Up

1. Clean the test tubes and place all glassware back in its proper location.
2. Wash your hands thoroughly before leaving the laboratory.

Analysis and Conclusions

Complete the **Analysis and Conclusions** section for this experiment either on your Report Sheet or in your lab report as directed by your teacher.

Part 1

1. The concentration of solution A is 0.02 M. How many moles of KIO_3 are in each milliliter of solution A?

2. Calculate the concentration of KIO_3 in moles per liter for each of the solutions after all the components have been mixed.

3. Plot a graph of the concentration-time data by plotting concentration of KIO_3 (or milliliters of A) on the vertical axis and plotting time on the horizontal axis. Connect the points with a smooth line.

4. Why is it important to keep the total volume at 10.0 mL during the dilutions of solution A?

5. What generalizations can you make concerning the effect on the time of the reaction resulting from varying the concentration of solution A?

6. How is the time of a reaction related to the rate of that reaction?

Part 2

1. Using your data, plot a graph of the temperature vs time (temperature on the vertical axis and time on the horizontal axis). Connect the points with a smooth line.

2. What general relationship can you derive from the graph?

3. Make a prediction of the time the reaction should have taken at 0°C and 75°C.

Something Extra

1. Use chemical resources to investigate other "clock reactions" in chemistry, such as the "Old Nassau" reaction.

2. Comment on the differences in the shapes of the two lines that you plotted.

Experiment 74

Equilibrium Beads

Problem
How can we model dynamic chemical equilibrium?

Introduction
A chemical system at equilibrium appears to be unchanging because the rates of the forward and reverse reactions are equal. However, such a system is dynamic because at the molecular level the forward and reverse reactions are constantly taking place. In this activity you will model a system achieving equilibrium.

Prelaboratory Assignment
✓ Read the entire experiment before you begin.
✓ Answer the Prelaboratory Questions.

   1. How will you tell when equilibrium is established?
   2. Why must the agitator be careful to always shake the box at the same rate?

Materials
Cardboard box reaction vessel
Blindfolds (2)
100 type A pop-it beads
50 type B pop-it beads

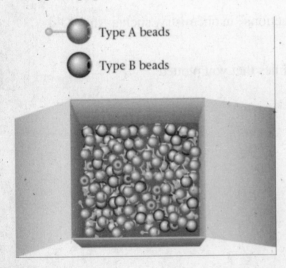

Type A beads

Type B beads

Safety
1. If any beads fall on the floor, pick them up immediately.

Procedure

Assign roles to the different members of your group:

a. Forward reaction -- the student finds two reactants and snaps them together to form a product molecule.

b. Reverse reaction -- the student searches for product molecules and unsnaps or breaks the bonds to form reactants.

c. Agitator -- the student agitates or shakes the cardboard reaction vessel to simulate the constant random kinetic motion of atoms and molecules.

d. Timer -- the student times the reaction and starts and stops the activity.

Part 1 $(A + A \rightarrow A_2)$:

1. Place 100 A type beads in the reaction vessel.

2. Blindfold the students representing the forward and reverse reactions.

3. The Agitator begins shaking the reaction vessel. Caution: the box needs to be shaken at a constant rate and care must be taken to not spill the beads.

4. The Timer signals the reactions to begin. The Timer stops the reaction after one minute.

5. Count the number of products (A_2) and determine the number of reactants. Record these numbers.

6. Leave all products intact and repeat the process until equilibrium is established, as shown by the same relative numbers of reactants and products for two or three consecutive 1-minute trials. Record the number of trials required to reach equilibrium.

Part 2 $(A + B \rightarrow AB)$:

1. All participants should switch roles.

2. Place 50 type A beads and 50 type B beads in the reaction vessel.

3. The Forward Reaction Student forms only the product AB. Repeat the procedure used for Part I. Record the numbers of "reactants" and products" in each trial and the number of times required to reach equilibrium.

Cleaning Up

Put all materials away as directed by your teacher.

Analysis and Conclusions

Complete the **Analysis and Conclusions** section for this experiment either on your Report Sheet or in your lab report as directed by your teacher. For each part you will need a Summary Table that includes the concentrations of reactants and products, as well as the value of the reaction quotient, Q, for each trial.

1. Calculate the value of "Q" for each trial in both Parts.

2. Compare your times required to reach equilibrium with those of other groups.

3. Compare your values of K with those determined by other groups.

4. Should you expect the same value of K for Part I and Part II? What causes differences in K values?

5. Should you expect the same value of K for each group in a given part? What causes differences in K values?

6. How does this activity show that equilibrium is microscopically dynamic but macroscopically static?

7. List some differences between this activity and a chemical system achieving equilibrium.

Something Extra

How would different numbers of beads affect the results? Carry out an experiment to answer this question.

Experiment 75

Equilibrium and Le Chatelier's Principle

Problem
How do systems at equilibrium respond to changing conditions?

Introduction
Le Chatelier's Principle describes the effect that applying various types of stress has on the position of a system at equilibrium...that is, whether it will shift to increase or decrease the concentration(s) of products in the equilibrium system. The Principle tells us that when a system at equilibrium is subjected to a stress, it will shift in such a way as to relieve the effects of that stress.

Stresses include variations in the concentrations of reactants or products, the temperature of the system, and (for reactions involving gases) the pressure. Of these, only a change in temperature actually changes the value of the equilibrium constant.

Most of our investigations occur in open systems, usually in water solution. Unless gases are involved in the reaction, the volume of the system is just the volume of the solution, and pressure is of little or no consequence. This permits us to simplify Le Chatelier's Principle to read: *For any reaction system at equilibrium in solution:*

> *if you add a reactant or a product to the system, it will try to consume what was added; if you remove a reactant or a product from the system, it will try to replace what was removed.*

In this experiment you will observe what Le Chatelier's Principle means. Your investigation will deal with two complex ions, both containing cobalt(II); they are $Co(H_2O)_6^{2+}$ and $CoCl_4^{2-}$. The procedure is short, and should not require more than 15 minutes to complete. When you have finished and cleaned your work area, return to your desk for the post-lab discussion, during which we will talk about what you saw and what it signifies in terms of the reaction system being investigated. You will need to take good notes during both the experiment and the post-lab session, since those notes will help you answer the questions pertaining to the experiment.

Prelaboratory Assignment
✓ Read the **Introduction** and **Procedure** before you begin.
✓ Answer the Prelaboratory Questions.
  1. The formula for solid cobalt(II) chloride is $CoCl_2 \cdot 6H_2O$. What name do we give to compounds which have water molecules bound to them?
  2. **a.** Write the equation for dissolving calcium chloride in water.
     b. Use Le Chatelier's Principle to predict the effect of addition solid calcium chloride to a solution containing both of the cobalt complexes.
  3. **a.** Write the equation for dissolving silver nitrate in water.
     b. Write the equation for the precipitation reaction that you would expect when a solution containing silver ions is added to a solution containing chloride ions.

Materials

Apparatus
50-mL Beaker
Shell vials, 1 dram (5)
Hot plate
Ice bath
Safety goggles
Lab apron

Reagents
Cobalt chloride hexahydrate, $CoCl_2·6H_2O$
Ethanol (or methanol)
12 M hydrochloric acid, HCl(aq)
Calcium chloride pellets, $CaCl_2(s)$
Acetone
Silver nitrate solution, $AgNO_3(aq)$, 0.10M

Safety

1. Safety goggles and a lab apron must be worn in the laboratory at all times.
2. Cobalt and silver solutions are mildly toxic, so you must wash your hands thoroughly before leaving the laboratory.
3. Silver nitrate will stain skin and clothing. Wipe up all spills with large amounts of water.
4. Concentrated hydrochloric acid will attack skin and clothing. Neutralize acid spills on laboratory surfaces before wiping up.
5. Be careful using the hot plate. Remember that hot surfaces look the same as cool ones. Use a hot pad to transfer hot containers from the hot plate.

Procedure   *An asterisk (*) indicates that written observations are called for in that step.*

1. Thoroughly dry your 50-ml beaker with a paper towel, then use the markings on the side to measure about 25-30 ml of ethanol into the beaker.

2. Examine the solid cobalt(II) chloride, noting both its color and the formula for the compound, as shown on the label of the stock bottle.*

3. Place a small sample of the solid (about the volume of two drops of water) in the beaker of ethanol.*

4. Divide most of the solution among five flat-bottomed vials, leaving about 0.5 cm of the solution in the beaker. The actual volume is not important, but they should all have approximately equal volumes of solution.

5. To one of the vials, add 5 drops of distilled water, one drop at a time, recording observations after each drop.* Duplicate the process with each of three other vials, so that all four are the same color. Use four of the five for Step 6, retaining one as a control, for comparison purposes.*

6. a. Take one of the vials (not the control) from Step 5 to the fume hood. Use the dropper provided with the acid to CAREFULLY add 5 drops of concentrated hydrochloric acid, one drop at a time, to the solution in the vial.*
   b. To a second vial from Step 5, add 2 small pellets of solid calcium chloride.*
   c. To the third vial, add 3-4 drops of acetone.*
   d. To your fourth vial, add 10 drops of 0.1M silver nitrate, $AgNO_3$, one drop at a time.*

7. To the solution remaining in the beaker, add just enough distilled water to get a color, about half-way between the blue and pink shades. This solution should now have approximately equal amounts of the two complex ions. Place the beaker on a hot plate and warm it until a color change occurs.*

8. Finally, chill the beaker in an ice bath, to see whether the color change in Step 7 is reversible.*

Cleaning Up

1. All solutions contain cobalt and should be transferred to the assigned waste container. The teacher will neutralize the acid (from step 6b) then dispose of the combined waste in an approved, safe manner.
2. Make certain that all glassware is clean, bottle caps are back in place, and that all is left as you found it.
3. Wash your hands with soap and water before leaving the laboratory.

Analysis and Conclusions

Complete the **Analysis and Conclusions** section for this experiment either on your Report Sheet or in your lab report as directed by your teacher.

The net-ionic equation for the equilibrium reaction you have been investigating is

$$Co(H_2O)_6^{2+}(aq) + 4Cl^-(aq) \rightleftharpoons CoCl_4^{2-}(aq) + 6H_2O(l)$$
$$\text{pink} \qquad\qquad\qquad\qquad \text{blue}$$

1. **a.** Which cobalt complex was favored by addition of water to the solution of cobalt(II) chloride in alcohol?
   b. Use Le Chatelier's Principle to explain the color change you observed.
2. **a.** Which cobalt complex was favored in both 6a and 6b?
   b. What ion is common to both of the reagents you used to bring about the color changes in these two steps?
   c. Use Le Chatelier's Principle to explain why the color changes occurred in each case.
3. Acetone absorbs water. Use this fact and Le Chatelier's Principle to explain the color change that you saw when you added acetone to the third vial in Step 6.
4. Silver chloride, AgCl, is a white solid. The equilibrium constant is $K = 6 \times 10^9$ for:

   $$Ag^+(aq) + Cl^-(aq) \rightleftharpoons AgCl(s)$$

   a. At equilibrium, would you expect to have mostly silver and chloride ions in solution, or mostly solid silver chloride? Explain.
   b. What color was the solid you formed in Step 6d? What must it have been?
   c. What color did the liquid in the vial turn? Which complex of cobalt was favored? Explain.
   d. Use Le Chatelier's Principle to explain why the liquid in the vial underwent the color change.
5. **a.** Which cobalt complex was favored by addition of energy as heat? Which complex was favored by cooling?
   b. Rewrite the equation for the reaction, including the energy term in the equation. The value of ΔH for the process is +50kJ/mol
   c. Use Le Chatelier's Principle and the equation from **5b** to explain the color changes that resulted from the heating and cooling.

Experiment 76

Chemical Equilibrium

Problem
Can an equilibrium constant be measured experimentally?

Introduction
Chemical equilibrium occurs in any closed system when the rate of the forward process is equal to the rate of the reverse process. When equilibrium is reached in a reaction mixture, macroscopic properties such as color remain constant. However, at the molecular level reactant molecules are becoming product molecules, and vice versa.

In this experiment you will examine the following reaction quantitatively:

$$Fe^{3+}(aq) + SCN^-(aq) \rightarrow FeSCN^{2+}$$

and determine the concentrations of each of the three ions at equilibrium. You will be able to do this because the iron(III) thiocyanate ion, $FeSCN^{2+}$, is colored.

The initial concentrations of the reacting ions will be calculated from dilution data, and the equilibrium concentration of the $FeSCN^{2+}$ will be determined by colorimetry. If you have ever looked carefully at a glass of colored liquid, such as cherry soda or iced tea, you know that the intensity of the color when viewed through the sides of the glass is much less than the color intensity when viewed from the top. This is because the color intensity depends upon the concentration of the colored substance *and* the amount of the solution you are looking through. In fact, a 1 cm length of a 1M colored solution will appear to have the same color intensity as a 2 cm length of a 0.5 M solution of the same material.

This experiment gives you the chance to compare five solutions of varying concentrations, and therefore, varying color intensity. You will compare them two at a time. You will adjust the relative depths of the two solutions until their color intensity appears to be the same.

Prelaboratory Assignment
✓ Read the **Introduction** and **Procedure** before you begin.
✓ Answer the Prelaboratory Questions.

1. Why is chemical equilibrium dynamic on a molecular level?

2. In a particular chemical reaction, solution A reacts with solution B to produce solution C. Solution A is colorless, solution B is light blue, and solution C is intensely pink. This reaction reaches equilibrium and gives a light purple color. How would you expect the color of this solution to change if you added more solution A? Explain your answer briefly.

3. In a particular reaction, X + 2Y = XY$_2$. If you begin with 3 M X and 4 M Y, and if at equilibrium 40% of the reactants have become products, how much of each species will you expect to find at equilibrium? What is the equilibrium constant for the reaction?

4. A solution of Fe(NO$_3$)$_3$ is 0.200 M. If you measure 10.0 mL of that solution and then dilute it to 25 mL with water, what is the resulting concentration of the Fe(NO$_3$)$_3$?

Materials

Apparatus
5 flat-bottomed vials or small test tubes
Disposable pipets
Diffuse light source
   (such as an overhead projector)
Test tube rack
Centimeter ruler
50-mL beaker
10-mL graduated cylinder
25-mL graduated cylinder
Paper or aluminum foil strip
Safety goggles
Lab apron
Pen or pencil for labeling

Reagents
0.00200 M potassium thiocyanate, KSCN
0.200 M iron(III) nitrate, Fe(NO$_3$)$_3$
Distilled water

Safety

1. Wear safety goggles and a lab apron at all times in the laboratory.
2. No food or drink is allowed in the laboratory at any time.
3. Be careful when using the light source. Do not stare directly into the light for extended periods of time.
4. Potassium thiocyanate is moderately toxic. Wash your hands thoroughly before leaving the laboratory.

Procedure

Part I   Preparing the Solutions

Note: Cleaning and drying glassware between each step of the procedure is important. Be careful not to contaminate solutions.

1. Line up five clean, dry flat-bottomed vials or small test tubes and label them 1,2,3,4, and 5.

2. Using your small graduated cylinder, measure 5.0 mL of 0.00200 M KSCN. Place it in vial 1. Repeat this with each of the four remaining vials.

3. Rinse your graduated cylinder well. Measure 5.0 mL of 0.200 M Fe(NO$_3$)$_3$ and place it in vial 1.
Note: Vial 1 will be used as the standard or reference vial.

4. Using your 25-mL graduated cylinder, measure 10.0 mL of the 0.200 M Fe(NO$_3$)$_3$ solution. Dilute it to exactly 25.0 mL with distilled water. Pour this solution into a clean, dry 50-mL beaker to mix it thoroughly.

5. Use 2-3 mL of the solution in the beaker to rinse your 10-mL graduated cylinder, then use the graduate to measure exactly 5.0 mL of the diluted solution. Pour it into vial 2. Save the remainder of the solution in the beaker for step 6.

6. Use 2-3 mL of the solution in the beaker to rinse your 25-mL graduated cylinder, then pour 10.0 mL of the solution from the beaker into the 25-mL graduated cylinder. Discard the rest as your teacher directs.

7. Fill the graduate from step 6 to the 25.0 mL mark with distilled water. As before, pour this solution into the *clean, dry* 50-mL beaker to mix it thoroughly.

8. After rinsing the 10-mL graduate with 2-3 mL of the solution in the beaker from step 7, use it to measure exactly 5.0 mL of this diluted solution and pour it into vial 3. Save the remainder of the solution in the beaker for step 9.

9. Rinse the 25-mL graduate with 2-3 mL of the solution in the beaker, then use it to measure exactly 10.0 mL of the solution. Discard the rest as your teacher directs.

10. Fill the graduate from step 9 to the 25.0 mL mark with distilled water. As before, pour this solution into the *clean, dry* 50-mL beaker to mix it thoroughly.

11. After rinsing the 10-mL graduate with 2-3 mL of the solution in the beaker from step 10, use it to measure exactly 5.0 mL of this diluted solution and pour it into vial 4. Save the remainder of the solution in the beaker for step 12.

12. Repeat steps 9-11, using your graduates and the solution remaining. **Note:** Discard the last remnant of solution in the small beaker. Each vial should have exactly 10.0 mL of solution in it: 5 mL of 2.00×10^{-3} M KSCN + 5 mL of $Fe(NO_3)_3$,, in 5 different concentrations.

Part 2   Comparison of Solutions
1. Find the light source that your teacher has instructed you to use. Take vials 1 and 2, along with your small beaker (cleaned and dried), a disposable pipet, and a piece of paper or foil to the light source.

2. Wrap the strip of paper or foil around vials 1 and 2 to block light from entering from the side.

3. Hold the two vials over the light source and look vertically down through the solutions toward the light.

4. Even if the color intensities appear the same initially, remove some of the solution from standard vial 1 (into the clean, dry beaker) with a pipet until the color intensity of the standard vial is definitely less than that of vial 2. Now add back some of the solution into vial 1 just until the color intensity in vial 1 matches that in vial 2.

5. When the color intensities are the same in both vials, return to your lab station. Measure the depth of the solution in both vials and record them.

6. Now repeat the color comparison with vials 1 and 3 varying the amount in vial 1 until it matches vial 3. Again, measure the depth of solution in both vials when the color intensities are the same. Record your results.

7. Continue to compare vial 1 with each of the successive vials vial 4 and vial 5.
   Note: You will only and always be removing and adding solution from or to vial 1.

Cleaning Up

1. Dispose of the solutions in all of the vials and beaker as instructed by your teacher.
2. Clean all glassware and return to its proper location.
3. Wash your hands before leaving the laboratory.

Analysis and Conclusions

Complete the **Analysis and Conclusions** section for this experiment either on your Report Sheet or in your lab report as directed by your teacher. Show a sample calculation for each of the questions 1-6, and then record your results in the Summary Table (obtained from your teacher).

1. Determine the initial concentration of Fe^{3+} in each vial.

2. Determine the initial concentration of SCN^- in each vial.

3. Determine the ratio of the depths of the solutions in the vials. Express this as the depth in Vial 1 divided by the depth in Vial n (the vial you're comparing in each trial).

4. Determine the equilibrium concentration of the $FeSCN^{2+}$ in each vial by multiplying the depth ratio for that vial by the concentration of $FeSCN^{2+}$ in vial 1.

5. Determine the equilibrium concentration of the Fe^{3+} in each vial. This is the initial Fe^{3+} concentration minus the amount that was converted to $FeSCN^{2+}$.

6. In similar fashion, determine the equilibrium concentration of the SCN^- in each vial.

7. Using expressions (a), (b), and (c) that are found at the bottom of the Summary Table, and using your calculated values for the equilibrium concentrations of $FeSCN^{2+}$, Fe^{3+}, and SCN^-, calculate the result for each "constant".

8. Only one of the expressions (a), (b), and (c) accurately represents the expression for an equilibrium constant. To determine which of those expressions resulted in the most "constant" constant, divide the largest answer in each column by the smallest answer in the column. The column that shows the number closest to 1 best represents the true equilibrium expression.

Experiment 77

MOM and Your CBL

Problem
How do liquid antacids work?

Introduction
Have you ever had to take an antacid such as milk of magnesia (MOM), Maalox™, or Gaviscon™? Pretty awful stuff: thick and gooey and no matter how they flavor it, it just doesn't taste good. How do liquid antacids work? This experiment will give you a chance to find out, while reviewing solubility concepts and acid-base chemistry.

Your stomach contains an acid mixture, roughly equivalent to 0.1 M hydrochloric acid, HCl(aq). The active ingredients in the antacids are primarily magnesium, and/or aluminum hydroxides, and carbonates. Among the other ingredients are compounds that help the hydroxide and carbonates adhere to the lining of your esophagus. As long as the acid remains in your stomach, there is no discomfort; it is only when the acid rises up into the esophagus that you feel the burning sensation. The solubility rules help you to understand why the products are so thick. They are suspensions of whatever active ingredients a particular brand employs. Due to the low solubilities of the hydroxides, there will only be a small amount of hydroxide ion present in solution at any given time.

Drinking the antacid introduces a slurry of metal hydroxide, only a portion of which is present as dissolve ions. The hydroxide ions in solution react immediately with hydrogen ions from the stomach acid. This causes more of the compound to dissolve so that more hydroxide ions are available for neutralization. The process continues either until the acid is neutralized or until all of the metal hydroxide is consumed.

In this experiment you will start with the antacid slurry and then add acid to it. To make the results easier to follow, you will use an acid solution that is much stronger than that in your stomach. The pH will be monitored with a CBL, using a pH probe.

Prelaboratory Assignment
   ✓ Read the **Introduction** and **Procedure** before you begin.
   ✓ Answer the Prelaboratory Questions.
   1. Why do the labels on these antacids always tell you to shake the bottle well before using?
   2. Write the dissociation equations for aluminum hydroxide and magnesium carbonate dissolving in water. Does the position of the equilibrium lie to the left or the right? Defend your choice.
   3. Using the appropriate equation, apply Le Chatelier's Principle to explain why a compound, such as magnesium hydroxide, which has very low solubility in water, will dissolve well in the presence of acid.

Materials

Apparatus

CBL, CBL2, or LabPro-
Graphing calculator, TI-83+ or similar
   with unit-to-unit connecting cable
Vernier pH probe and amplifier
Application program:
   DataMate for LabPro or CBL2, ChemBio for original CBL
Beaker, 250 mL
Magnetic stirrer and stirring bar*
Plastic syringe, 3-5 mL
Safety goggles
Lab apron

Reagents

10% Hydrochloric acid, HCl(*aq*)
Liquid antacids
   (Maalox, Milk of Magnesia, Gaviscon, etc.)

* If these are not available, you will need a glass stirring rod. In addition, a ring clamp is recommended as a protective collar, to prevent the beaker from tipping over.

Safety

1. Hydrochloric acid is corrosive to skin and clothing. If it is spilled on the desk top, wipe it up with large amounts of water. Large spills should be neutralized with baking soda before they are wiped up. Skin should be flooded with water for several minutes. In the case of clothing spills, remove the affected article, and wash thoroughly with water. Report any spill to your teacher immediately.
2. Safety goggles and a lab apron must be worn at all times in the laboratory.

Procedure

If you are not using a Report sheet for this experiment, design a Data Table with the following headings:

Trial	Initial pH	Minimum pH	Time to min. pH	10-Second pH	Final pH	Remarks

CBL Set-Up. Following the instructions for your graphing calculator, prepare the CBL to collect pH readings (ChemBio for TI-83+; the PH program for other calculator models). Choose Set Up Probes from the Main Menu. Make the appropriate selections for 1 probe in Channel 1, to measure pH. When the calculator returns you to the Main Menu, select option 2: Collect Data, followed by Time Graph. You will be asked for Time Between Samples: 0.1 [ENTER]; then for the number of samples: 200 [ENTER]. The next screen will give you a summary of your selections; [ENTER]. Next, you will be asked if you want to keep the time set up; you do – [ENTER]. When the calculator asks you for Y Minimum and Y Maximum, select 1.0 and 12.0, respectively. For Yscl, enter 0.1. This will allow you to collect your data as a function of time, and will give you a total run time of 20 seconds per run; you will conduct at least 5 runs.

1. A sketch of the apparatus is shown **Figure 1**. Use a wooden, clothes pin-type clamp to hold the pH probe in place.

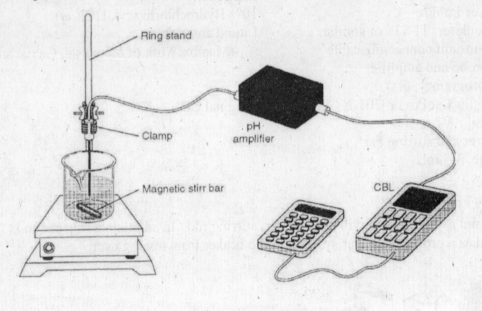

Figure 1
Apparatus for Determining pH with CBL

2. Connect the pH amplifier and pH probe to the CBL.

3. Place 20 mL of liquid antacid in the beaker, with 100 mL of distilled water. Place the stirring bar in the beaker with the mixture. Be sure the tip of the pH probe is covered by the antacid-water mixture, but is high enough that the stirring bar will not hit it. If necessary, add distilled water to the beaker to give added depth. Turn on the stirrer and adjust the speed to as high as possible without causing a vortex (whirlpool). You are now ready to begin. Recall that each complete run of the experiment is only 20 seconds long, so be sure both partners are ready.

4. As one partner presses [ENTER], the other waits for one second (estimated) then injects 2.0 mL of 10% hydrochloric acid into the beaker. When the 20 seconds have expired, the CBL screen will read, **Done**.

   Note: If stirring is done manually, the person adding the acid should also press [ENTER].

5. The calculator will tell you where the time and pH data are stored (probably L_1 and L_2). Accept this by pressing [ENTER]. The calculator will then display the graph that was produced during the experiment.

   When you pressed [ENTER], the pH reading dropped briefly to (about) 7. This is a false reading, and should be ignored. Use the right arrow on your calculator to trace along the curve until you reach the highest pH that precedes the sharp drop. This is taken to be the pH at the instant that you added the hydrochloric acid, and is recorded under "Initial pH" in the data table.

   Continue tracing along the curve until you get to the lowest pH value. Record this value and the number of readings between the initial and lowest pH values. Since each new reading represents 0.1s, you can determine the time elapsed during the sharp drop in pH.

Finally, trace all the way to the end of the curve to get the highest pH reached in the 20-second run. Record this value and the current pH reading (by the time you take this pH reading, the value displayed by the CBL unit can be assumed to be "final").

6. Repeat the experiment until you have 5 trials, or until the mixture in the beaker turns clear.

Cleaning Up

1. Check the CBL reading to determine the pH. If it is between 6 and 8, the mixture is sufficiently neutral and it can be flushed down the drain with water. Be careful not to pour out the magnetic stirring bar. If the pH is outside the allowable range, adjust it as necessary, using acid and antacid from the experiment.
2. Wash your hands thoroughly before leaving the laboratory.

Analysis and Conclusions

1. Describe the shapes of the graphs you obtained. Offer an explanation for the observed decrease in pH, followed by a return to nearly the original level. In other words, what caused the decrease, and why does the pH then increase again?

2. Notice that the pH drops very rapidly, but returns slowly to a value near the initial one (except, perhaps, for the final trial).
   a. Addition of acid to a slurry of milk of magnesia causes a sharp reduction in pH; why?
   b. The return to higher pH is much slower than the initial drop. Why? What must happen for the pH to rise? (**Hint:** Two things must occur.)

3. It is likely that the mixture in the beaker turned clear after the last trial. What does that signify about the nature of the mixture remaining in the beaker?

4. Examine the values in the "Final pH" column. Account for the consistency of the values. (**Hint:** What did the mixture look like after each trial?)

5. Gram for gram, which would you expect to be the most effective antacid: magnesium hydroxide or aluminum hydroxide? (**Hint:** How many moles of hydroxide ion is present in one gram of each compound?)

6. Gaviscon™ uses both aluminum hydroxide and magnesium carbonate. How do these two compare, gram for gram, as neutralizers?

7. Look up values for the solubility product constants, K_{sp}, for magnesium hydroxide, aluminum hydroxide, and magnesium carbonate. Suppose you were to repeat the procedure using just one of the three compounds, rather than the commercial products. Speculate as to how the relative K_{sp} values might influence the shape of the pH vs. time curves you made.

Something Extra

1. Repeat the experiment using a different antacid. Compare the two sets of results.

2. Compare name brands (such as Phillips, Maalox) with store brands containing the same active ingredients.

3. Test your predictions from question **7**, of **Analysis and Conclusions**.

Experiment 78

Oxidation and Reduction

Problem

What kinds of reactions involve electron transfer as the driving force?

Introduction

Many reactions involve a transfer of electrons from one atom (or ion) to another. Such processes are known as *oxidations* (for electron loss), or *reductions* (for electron gain). Since they always occur together, they are often referred to collectively as "*redox reactions*."

If an atom loses electrons, those electrons must be transferred to some other atom or ion; the atom or ion that takes the electrons is called the *oxidizing agent*. This name suggests that the receiver of the electrons caused the oxidation to occur. Similarly, the species that loses the electrons is called a *reducing agent*, because it provided the electrons that caused the other atom or ion to become reduced.

In this experiment, you will learn about some of the types of reactions in which electron transfer is the driving force. You have seen many of these before and you may know them under different names. The first three parts of the experiment will be carried out in a 96-well test plate. Parts 4 and 5 involve reagents that can attack the plastic, so you will use small test tubes for those parts.

Prelaboratory Assignment

✓ Read the **Introduction** and **Procedure** before you begin.
✓ Answer the Prelaboratory Questions.

1. **a.** What are the most common ionic charges for the elements shown in their ionic compounds?

   Aluminum   Oxygen

   b. Complete the following equations showing aluminum metal and oxygen gas forming the ions you identified in **1a**.

   i. $Al(s)$ $\rightarrow$ $+$

   ii. $O_2(g) +$ $\rightarrow$ 2

   c. Identify each of the equations in **1b** as oxidation or reduction.

2. Write the balanced equation for the reaction between aluminum metal and oxygen gas to form aluminum oxide.

3. Identify each of the following processes as an example of oxidation, reduction, or both.
   a. Chlorine gas forms chloride ion.
   b. Sodium metal forms sodium ions.
   c. Iron rusts.
   d. Hydrogen peroxide breaks down into water and oxygen gas. (**Hint:** the oxygen in hydrogen peroxide, H_2O_2, has an oxidation state of -1.)
   e. Household bleach removes a stain from an article of clothing.

Materials

Apparatus
96-well test plate
Wash bottle
24-well plate to hold reagent pipets
Forceps or tweezers
Toothpicks as stirring rods
75-mm test tubes (5)
Safety goggles
Lab apron

Solid Reagents
Magnesium foil (small pieces)
Copper foil (small pieces)
Zinc foil (small pieces)

Reagent Solutions (In thin-stem pipets)
1.0 M HCl(aq)
1.0 M Cu(NO$_3$)$_2$(aq)
1.0 M Zn(NO$_3$)$_2$ (aq)
0.10 M KMnO$_4$(aq)
0.10 M FeCl$_3$ (aq)
0.10 M FeSO$_4$(aq)
0.50 M NH$_4$VO$_3$ (aq)
0.1 M oxalic acid, H$_2$C$_2$O$_4$
0.5 M Na$_2$SO$_3$(aq)
3% H$_2$O$_2$ (household antiseptic solution)
Chlorine water, Cl$_2$(aq)
Bromine water, Br$_2$(aq)
Iodine water, I$_2$(aq)
0.2 M NaI(aq)
0.2 M NaBr(aq)
3 M H$_2$SO$_4$(aq)
Bleach, 3% aqueous NaOCl

Safety

1. Safety goggles and a lab apron must be worn at all times in the laboratory.
2. Hydrochloric, sulfuric and oxalic acids are corrosive to skin and clothing.
3. Wipe up all spills immediately, using large quantities of water.
4. Small quantities of gaseous halogens may be generated. Work in a well-ventilated area.

Procedure

Part 1

1. Place 10 drops of 1.0 M HCl in a well of your test plate. Add a small piece of magnesium then observe and record what happens as the reaction proceeds.

Part 2

2. Add 10 drops of 1.0 M copper nitrate, Cu(NO$_3$)$_2$, to one well of your test plate and add a small piece of zinc metal. Record your observations, both of the appearance of the metal and of the solution color. Changes may not happen right away.

3. Put 10 drops of 1.0 M zinc nitrate, Zn(NO$_3$)$_2$, in a different well of your test plate. Add a small piece of copper metal. Observe and record any reactions which take place, including solution color and changes at the surface of the metal.

Part 3

4. Place 10 drops of 0.1 M iron(III) chloride (a source of Fe^{3+} ion) in one well of your test plate and 10 drops of iron(II) sulfate (a source of Fe^{2+} ion) in another well. Add 1 drop of 0.1 M potassium permanganate, KMnO$_4$, to each well. One of the wells will show a change in the color of the KMnO$_4$; the other will not. Note which iron ion reacts (decolorizes) with the KMnO$_4$.

5. To the well from step 4 in which a color change was seen, continue adding potassium permanganate until the purple color of the permanganate ion no longer changes. Add one drop at a time, stir after each drop, and record the number of drops needed to complete the reaction.

Part 4. The reagents for this part of the experiment will attack the plastic well plate, so small test tubes and Erlenmeyer flasks will be used.

6. Place 5 drops of sodium iodide solution, NaI, in each of two test tubes. To the first, add two drops of chlorine water (chlorine gas dissolved in water). To the second, add two drops of bromine water. In each case, look for a color change, indicating that a chemical reaction has occurred.

7. Using clean tubes, place 5 drops of sodium bromide solution, NaBr, in each of two small test tubes. Add two drops of chlorine water to one, and two drops of iodine solution to the other. As before, look for evidence of a chemical reaction, in the form of a color change.

8. For many transition metals, more than one possible cation can be formed. When an atom or ion changes its charge, we say it has undergone a change in oxidation state. Since these oxidation state changes are often accompanied by a change in the color the metal ion gives to aqueous solutions, it is a simple matter to tell when an oxidation or reduction of the metal occurs.
   a. Manganese in the +7 oxidation state in potassium permanganate, $KMnO_4$.
      i. Put 10 drops of hydrogen peroxide antiseptic in a small tube. Add 1 drop of 3 M H_2SO_4 and tap the side of the tube gently to mix the reagents. Now add 1 drop of $KMnO_4$, potassium permanganate, solution and observe the changes that take place. (Hint: you should see two changes, only one of which involves the color of the $KMnO_4$.)
      ii. Repeat the previous step, but omit the sulfuric acid.
      iii. Repeat the experiment, but this time use 0.1 M oxalic acid solution, $H_2C_2O_4$, in place of the hydrogen peroxide and sulfuric acid. If no changes are evident after 1 minute, try warming the tube on a hot plate.
   b. Vanadium in the +4 and +5 oxidation states.
      i. Place about 1 mL of ammonium vanadate solution in a 10-mL Erlenmeyer flask. The color you see is due to vanadium in the +5 state. Add an equal volume of sodium sulfite solution and swirl the flask. If no change occurs within 30 seconds, heat the flask gently on the hot plate. Note and record the color of the solution, which now contains vanadium in the +4 state.
      ii. Allow the flask from the previous step to cool, then add a few drops of bleach, a 3% solution of sodium hypochlorite, NaOCl(*aq*). Observe and record any changes.

Cleaning Up

1. Place any bits of unreacted metal that may remain in your test plate in the appropriate waste container.
2. To the tubes from **Part 3**, steps 6 and 7, add about 1 mL of 0.5 M sodium sulfite, solution, Na_2SO_3, to convert any remaining molecular halogens to the corresponding halide ions.
3. Flush the remaining solutions down the drain with large amounts of water. While some of the solutes are potential hazards and others are quite acidic, the dilution provided by a running stream of water will bring the concentrations to acceptable levels.
4. Return the pipets and unused solutions to their proper places. Do not empty them into the sink.
5. Wash your hands thoroughly before leaving the laboratory.

Analysis and Conclusions

Complete the **Analysis and Conclusions** section for this experiment either on your Report Sheet or in your lab report as directed by your teacher.

The list below shows six equations which you will need in order to answer the questions for **Parts 1-3**. They are called half-reactions because each shows only half of an oxidation -reduction process. They are written twice: first as reductions, then reversed to become oxidations.

$$Zn^{2+} + 2e^- \rightarrow Zn \qquad\qquad Zn \rightarrow Zn^{2+} + 2e^-$$
$$Cu^{2+} + 2e^- \rightarrow Cu \qquad\qquad Cu \rightarrow Cu^{2+} + 2e^-$$
$$Mg^{2+} + 2e^- \rightarrow Mg \qquad\qquad Mg \rightarrow Mg^{2+} + 2e^-$$
$$Fe^{3+} + 1e^- \rightarrow Fe^{2+} \qquad\qquad Fe^{2+} \rightarrow Fe^{3+} + 1e^-$$
$$2H^+ + 2e^- \rightarrow H_2 \qquad\qquad H_2 \rightarrow 2H^+ + 2e^-$$
$$Cl_2 + 2e^- \rightarrow 2\,Cl^- \qquad\qquad 2\,Cl^- \rightarrow Cl_2 + 2e^-$$

Part 1

1. The only possible states for magnesium are the neutral element (zero charge) and the positive ion, Mg^{2+}.
   a. What was the initial oxidation state of magnesium?
   b. What was the oxidation state of magnesium after it reacted with HCl?
   c. Was magnesium oxidized or reduced? Explain.
   d. Write the half-reaction showing the change for magnesium.

2. The hydrochloric acid solution contains H^+ ions and Cl^- ions. Consult the list of half-reactions to decide what gas was formed in the reaction. (**Hint:** Recall that you need one oxidation and one reduction taking place in order to have a complete reaction, so this gas must form as a result of the choice you did *not* select in **1c**, above.) Write the equation for the half-reaction resulting in the evolution of the gas you select.

3. Combine your half-reactions from questions **1** and **2** to write the equation for the complete reaction.

Part 2

4. Consider the two wells used for zinc and copper.
   a. What evidence of reaction did you see, if any?
   b. One of the following equations represents the reaction that occurred. Select the correct one and justify your choice.

   $$Cu + Zn^{2+} \rightarrow Cu^{2+} + Zn \qquad or \qquad Cu^{2+} + Zn \rightarrow Cu + Zn^{2+}$$

   c. For the reaction equation you selected in **4b**, identify each of the following:
   i. the species reduced   ii. the species oxidized
   iii. the reducing agent   iv. the oxidizing agent

Part 3

5. **a.** In which well did the color of permanganate ion disappear? That is, which test reagent, Fe^{2+} or Fe^{3+}, caused the purple color to fade?

   b. The permanganate and iron solutions were both 0.1 *M* yet ten drops of iron solution were able to discolor only a few drops of permanganate solution. What can you conclude about the stoichiometry of the reaction?

6. The color change was the result of one of the iron ions being converted to the other oxidation state.

   a. Which of the two equations below best illustrates what happened in the well for which you observed the color change? Explain how you arrived at your choice.

$$Fe^{2+} \rightarrow Fe^{3+} + e^- \qquad \text{or} \qquad Fe^{3+} + e^- \rightarrow Fe^{2+}$$

   b. Is iron being oxidized or reduced? Explain.

   c. Is permanganate ion being oxidized or reduced? Explain.

   d. In the reaction between permanganate and iron ions, which is the oxidizing agent and which is the reducing agent? Explain.

Part 4

7. The color you saw in step 6 is due to formation of molecular iodine, I_2, from iodide ion, I^-.

   a. Is iodine being oxidized or reduced in this reaction?

   b. Oxidation and reduction must always occur together. What species in each of the tubes must be causing the change from iodide to iodine?

8. In **step 7**, only one of the halogen solutions was able to change bromide ion, Br^-, to molecular bromine, Br_2. Which halogen caused the change?

9. Look at your combined results from steps 6 and 7, and consider the relative positions of chlorine, bromine and iodine on the periodic table. Make a general statement concerning the relative ease of oxidizing halogen ions to molecular halogens.

Part 5

10. For the reaction between hydrogen peroxide and potassium permanganate:

    a. What difference does the presence of sulfuric acid make?

    b. What is the gas formed in the reaction?

11. How does the speed of reaction between permanganate and oxalic acid compare with the speed of most of the other reactions in this experiment? What was the effect of warming the system on the speed of reaction?

12. Did the products of the reaction between potassium permanganate and oxalic acid more closely resemble those of the reaction between hydrogen peroxide with sulfuric acid or without? Account for this result.

13. In the reaction between $KMnO_4$ and $H_2C_2O_4$, a gas was produced as the oxalic acid reacted. What was this gas? (**Hint:** Look at the relative proportions of the elements in oxalic acid.)

14. The sulfite ion changes vanadium from the +5 state to the +4 state.

    a. Is this an oxidation or a reduction?

    b. What must have happened to the sulfite ion in the process?

15. What effect (oxidation or reduction) does the hypochlorite ion in $NaOCl(aq)$ have on vanadium(IV)? What happens to the vanadium(IV)? How can you tell?

Experiment 79

Activity Series

Problem
Which metals are the most active? Which are the least active? How do we measure activity?

Introduction
In this lab you will react various metals with different solutions and rank the activities of the metals.

Prelaboratory Assignment
✓ Read the entire experiment before you begin.

Materials
Safety goggles
Lab apron
Well plates (24 wells)
Tweezers or forceps

Metal strips (Cu, Al, Zn, Fe, and Mg)
HCl (6.0 M)
0.10 M solutions of
    $CuCl_2$,
    $Fe(NO_3)_2$
    $Zn(NO_3)_2$
    $Mg(NO_3)_2$
    $Al(NO_3)_3$

Safety
1. Safety goggles and a lab apron must be worn at all times in the laboratory.
2. The 6.0 M HCl is corrosive. Handle it with extreme care.
3. If you come in contact with any solution, wash the contacted area thoroughly. Wipe up any spills immediately.

Procedure
Part 1   Developing an Activity Series
1. Place a few drops of one solution to wells 1-5 in Row A of your well plate. Put drops of a second solution in wells 1-5 of Row B, and so on.

2. Place a strip of one of the four of the metals to each solution in Row A, a second metal to Row B, and so on.

3. Record all changes observed in the metals and solutions. Clean the well plate as described in steps 1 and 2 of **Cleaning Up**, then test each of the five solutions with the remaining metal. Record your observations for the fifth metal, then proceed to **Part 2**.

Part 2   Testing the Activity Series

1. Place a few drops of 6.0 *M* HCl in 5 separate reaction wells.

2. Place a strip of each metal in separate solutions of 6.0 *M* HCl.

3. Record all changes observed in the metals and solutions.

Cleaning Up

1. Use a tweezers or forceps to remove metals from the wells. Place the metals as your teacher directs.
2. Rinse the wells thoroughly with water. Dispose of the solutions as directed by your teacher.
3. Place the well plate upside down on a paper towel to drain.
4. Wash your hands thoroughly before leaving the laboratory.

Analysis and Conclusions

Complete the **Analysis and Conclusions** section for this experiment either on your Report Sheet or in your lab report as directed by your teacher.

Part 1

1. Suppose metal "A" reacts with a solution containing ions of metal "B". Can we predict whether or not metal "B" will react with a solution containing ions of metal "A"?

2. According to your results, order the metals from most active to least active. Explain your reasoning. This order is called the activity series.

Part 2

3. Are your results in **Part 2** consistent with the activity series you developed in **Part 1**? Explain. Where would H^+ fit in this series?

4. Identify the oxidizing agent and reducing agent in each reaction.

5. Do metals react with solutions containing ions of the same metals? Explain the results when a metal was added to a solution containing the same metal.

Something Extra

Would changing the concentrations of the solutions affect your activity series? Try it.

Experiment 80

Halogen Activity Series

Problem

How do the halogens compare in terms of their relative ease of reduction to their corresponding halide ions?

Introduction

Elements are described as being *active* if they tend to react with other elements to form compounds readily; in that sense, the word active really means "reactive." In contrast to metals, which can only form ionic bonds, nonmetals can form compounds either by ionic bonding or by covalent bonding. When they bond ionically, nonmetals tend to gain electrons, forming anions. Covalent bonding results from the sharing of electrons, as the atoms share electrons in a way that fills the available orbitals in each. Thus, an active nonmetal is one that has a strong tendency to acquire electrons.

Just as oil and water don't mix in salad dressings, many organic liquids are not able to mix with water. These liquids are referred to as being nonpolar. You first learned about polar and nonpolar molecules in Chapter 12. The nonpolar liquid you will use in this experiment is ordinary paint thinner. While paint thinner is a mixture of nonpolar solvents, it will do nicely for this purpose.

As you might recall from Chapter 15, substances such as table salt which dissolve well in water are not very soluble in nonpolar liquids, while those substances that don't dissolve appreciably in water are often quite soluble in nonpolar liquids. In this experiment, you will take advantage of the differences in solubility between the halogens in molecular form (F_2, Cl_2, etc.) and the corresponding halide ions (F^-, Cl^-, Br^- and I^-). Their charges make the ions very soluble in water, while the molecular halogens are only slightly soluble. On the other hand, the nonpolar molecular halogens are very soluble in a nonpolar liquid, but only sparingly so in water.

By placing molecular halogens in direct competition for electrons with the halide ions, you will be able to compare their relative electron-attracting abilities. Based on tests with certain of the combinations, you should be able to infer the order of relative activities for the entire halogen family.

Prelaboratory Assignment

✓ Read the **Introduction** and **Procedure** before you begin.
✓ Answer the Prelaboratory Questions.

1. Write equations for the formation of the halide ions from molecular halogens. Assume the reactions take place in aqueous solution.

2. Do the equations you wrote represent oxidations or reductions? Defend your choice.

3. **a.** Write the equation for the single replacement reaction between aqueous sodium chloride and molecular fluorine.

   b. Which species is oxidized? Which is reduced? Answer using names and symbols of the species involved.

4. Explain why halide ions dissolve well in water, while molecular halogens do not.

5. Steps 2, 3, and 4 of the **Procedure** each exclude one of the halide ion solutions. Why? (**Hint:** Would chlorine atoms be likely to react with chloride ions by taking away the extra electron?)

Materials

Apparatus
13 ×100-mm test tubes
     (at least 3, preferably 6, optimum 12)
24-well test plate or other rack for test tubes
Corks or stoppers for tubes
Safety goggles
Lab apron

Reagents
Paint thinner
Iodine solution, I_2 (aq)
Bromine solution, Br_2 (aq)
chlorine solution, Cl_2 (aq)
0.2 M NaF(aq)
0.2 M NaCl(aq)
0.2 M NaBr(aq)
0.2 M NaI(aq)
$Na_2S_2O_3$(aq), saturated (for cleanup)

Safety

1. The aqueous halogen solutions are toxic and have strong, unpleasant odors. Work only in well-ventilated areas.
2. Paint thinner is highly flammable. There can be no open flames in the laboratory.
3. Wear safety goggles and lab aprons at all times in the laboratory.

Procedure

Part 1   Preliminary Tests

1. Place 0.5 mL of paint thinner in each of three test tubes. To one of the tubes add 0.5 mL of chlorine water, Cl_2 (aq); add 0.5 mL of bromine water, Br_2 (aq) to the second; and place 0.5 mL of iodine water, I_2 (aq), in the third. Cork the tubes and shake briefly to mix. Note and record the appearance of the contents of each tube. Record your observations.

Part 2   Activity Tests

1. Into each of three separate test tubes, place 0.5 mL of one of the three halide ion solutions: NaF, NaBr, and NaI. Be sure to keep track of which halide ion is in which tube. Add 0.5 mL of paint thinner to each tube, followed by 0.25 mL of chlorine water. Cork the tubes and shake briefly to mix. Allow the contents to settle and make note of the appearance of both layers in each tube.

2. Repeat the process, this time substituting bromine water for chlorine water and using NaCl in place of NaBr. Describe the results of the tests on fluoride ion, chloride ion, and iodide ion. If you need to reuse the tubes, be sure to clean them out as described below under **Cleaning Up**.

3. Finally, test solutions of fluoride, chloride, and bromide ions with iodine solution, record the results, and complete the **Cleaning Up** procedure.

Cleaning Up

While the halide ion solutions present no chemical hazard, both paint thinner and the molecular halogens must be disposed of with caution.

1. Add about 1 mL of sodium sulfite solution to each tube. Cork the tube and gently shake it. This will convert the molecular halogens to halide ions and will move them out of the organic layer (paint thinner) and into the lower aqueous layer. Consult your instructor concerning disposal from there.
2. Use a small brush and warm, soapy water to clean the tubes. Return them to their proper locations.
3. Wash your hands thoroughly before leaving the laboratory.

Analysis and Conclusions

Complete the **Analysis and Conclusions** section for this experiment either on your Report Sheet or in your lab report as directed by your teacher.

1. Write equations for the reactions that took place. You need not write equations for any cases in which there was no evidence of chemical reaction.

2. Examination of your data should let you place bromine, chlorine, and iodine in order of activity. A more active molecular halogen is able to oxidize (take electrons) from less-active halide ions. Once you have established the order for these three, you should be able to place fluorine relative to the others, based on their positions on the periodic table.

3. Suppose you had been provided with a solution of molecular fluorine, F_2 (aq). Predict the reactions that you would expect as F_2 (aq) is mixed with paint thinner and solutions of the other halides.

4. It would be impossible for your instructor to prepare an aqueous solution of molecular fluorine. Suggest an explanation for this fact.

Experiment 81

Analysis of Hydrogen Peroxide
A Gravimetric Titration

Problem

Does a commercial antiseptic have the full potency claimed on the label?

Introduction

The label on a bottle of commercial antiseptic reads, "3 Percent Hydrogen Peroxide." Is this true? How precisely is the concentration known? Hydrogen peroxide spontaneously decomposes into water and oxygen gas, so the actual concentration is constantly decreasing. The percentage of hydrogen peroxide, H_2O_2, in the antiseptic at any given time can be determined by testing it with potassium permanganate solution. In the reaction, as in the spontaneous decomposition, part of the oxygen is converted (*oxidized*) to oxygen gas, O_2, while the hydrogen and the remaining oxygen atoms become water. The manganese in the permanganate ion is converted to manganese (II) ions, Mn^{2+}; the other products are shown in the equations below. The molecular equation for the reaction is

$$2\ KMnO_4 + 5\ H_2O_2 + 3\ H_2SO_4 \rightarrow 2\ MnSO_4 + 5\ O_2 + K_2SO_4 + 8\ H_2O$$

and the net ionic equation (which shows only those ions and molecules which are actually involved in the reaction) is

$$2\ MnO_4^-\ (aq) + 6\ H^+(aq) + 5\ H_2O_2\ (aq) \rightarrow 5\ O_2(g) + 2\ Mn^{2+}(aq) + 8\ H_2O(l)$$

While the chemistry is complex the procedure is a titration similar to the ones used earlier to analyze for the acidity of vinegar or the amount of calcium in skim milk. In this case, the titrant is potassium permanganate solution, the concentration of which will be given in *moles of solute per gram of solution* . From the mass of permanganate solution used you will be able to find the number of moles of permanganate ions that have reacted. From the reaction equation, which shows that permanganate ions and hydrogen peroxide react in a 2:5 ratio, you will be able to determine the number of moles of hydrogen peroxide that react. From these data you can determine the percentage of H_2O_2 in a bottle of commercial antiseptic. If you wish, you can also investigate the effect that temperature and/or aging has on the potency of the peroxide solution.

You will use three solutions, each in its own pipet: potassium permanganate, $KMnO_4$; hydrogen peroxide, H_2O_2; and 6 *M* sulfuric acid, H_2SO_4. You will need to keep track of the masses of the permanganate and peroxide solutions, but not of the acid.

Prelaboratory Assignment

✓ Read the **Introduction** and **Procedure** before you begin.
✓ Answer the Prelaboratory Questions.

1. If antiseptic hydrogen peroxide is 3 % H_2O_2 by mass, approximately how many moles of H_2O_2 are present in each gram of antiseptic solution? You can assume the solution has the same density as water. Show your calculations.

2. A solution of potassium permanganate is prepared by dissolving 1.45 g of solid $KMnO_4$ in about 100 mL of water. The total mass of solution is 102.50 g. Assuming the density of the solution is 1.0 g/mL, calculate the concentration of the solution:
   a. in moles of solute per gram of solution. Show your work.
   b. in moles of solute per liter of solution (the molarity). Show your work.

3. Your three titrations will require a total of about 0.30 mL of 6 M H_2SO_4. What mass of sodium bicarbonate would be needed to neutralize this much acid? The molecular equation for the neutralization reaction is

$$2\,NaHCO_3(s) + H_2SO_4\,(aq) \rightarrow 2\,H_2O(l) + 2\,CO_2(g) + Na_2SO_4(aq)$$

Materials

Apparatus
Milligram balance
10-mL Erlenmeyer flasks (3)
pH test paper
Safety goggles
Lab apron

Reagents
$KMnO_4$ solution in microtip pipet
3% H_2O_2, (commercial antiseptic)
6M H_2SO_4 in thin-stem pipet
Baking soda, $NaHCO_3$, for clean-up

Safety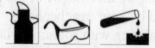

1. **This experiment involves two potentially hazardous materials:** potassium permanganate, $KMnO_4$, and 6M sulfuric acid, H_2SO_4. The first is a strong oxidizing agent, while the second is a highly corrosive acid. Both will react quickly with skin and clothing. Wash off a spill of either solution with large amounts of water. Notify your teacher at once of any spills.
2. Safety goggles and a lab apron must be worn at all times in the laboratory.

Procedure

For each of your three titration samples, record the pipet masses and determine the mass of solution used.

1. Fill a thin-stem pipet with commercial hydrogen peroxide solution and label the pipet accordingly. Fill a second pipet with $KMnO_4$ solution (dark purple), taking care to keep the solution off your skin. Record the concentration of the solution as given on the stock bottle of permanganate; you will need this for your calculations. Label the pipet containing the permanganate solution. Weigh both filled pipets and record their masses to the full precision of the balance.

2. Add 20 drops of the hydrogen peroxide solution to a 10-mL Erlenmeyer flask, then add 3 drops of 6M sulfuric acid, H_2SO_4. Weigh the pipet with the H_2O_2 solution and record the mass.

3. Add $KMnO_4$ solution dropwise to the flask, a few drops at a time, and swirl the contents to mix the solutions. If a drop of the permanganate clings to the side, use a drop of the acid solution to rinse it down into the rest of the liquid, otherwise it will react with the glass surface, causing a brown stain. These stains result from a side-reaction which can affect your results adversely. Continue adding the permanganate and mixing the solutions until a faint pink color persists.

4. Determine and record the mass of the permanganate pipet. As soon as you have finished this trial, clean your flask as directed below, under **Cleaning Up**; the products will stain the titration vessel if left for more than a minute or so.

5. Repeat the experiment twice more, for a total of three trials. Use the same pipets, refilling them as needed. Start with a clean flask each time.

Cleaning Up

1. The solution in your vessel at the end of each titration is highly acidic, and must be neutralized before you dispose of it. Pour the contents of the flask into a beaker (100-mL or larger) and add a small amount of baking soda or a little saturated sodium carbonate solution, a mild base. The solution may foam up as the carbonate solution reacts with the acid. The gas escaping is carbon dioxide, CO_2. Gentle swirling of the beaker speeds the neutralization. Continue adding small amounts of base until no further foaming occurs. Leave the beaker at this point until you have completed your final trial.

2. Once the acid from all of your titrations has been neutralized, test the pH of the mixture by using a glass rod to place a drop on a piece of pH test paper. Compare the color of the drop on the paper with the color chart that is supplied with the paper. If it shows that the pH of the mixture is between 5 and 9, pour the liquid into the container designated by your teacher as Manganese(II) waste. If the pH is still below 5, continue neutralizing with sodium carbonate solution until a safe pH value is reached. Because Mn^{2+} is somewhat toxic, the solution will be treated further before disposal.

3. Place your permanganate and acid pipets, with any remaining solution, in the place designated by your teacher. **Do not** empty them into the sink or return the unused solutions to the stock bottles. Your teacher will tell you if the peroxide pipet is to be used again; if it is not, the contents may be rinsed down the drain with water.

4. Wash your hands thoroughly with soap and water before leaving the laboratory.

Analysis and Conclusion

Complete the **Analysis and Conclusions** section for this experiment either on your Report Sheet or in your lab report as directed by your teacher.

If you are not using a Report sheet for this experiment design a Summary Table which includes the following items for each of your samples:
- ✓ mass of $KMnO_4$ solution used
- ✓ moles of $KMnO_4$ used
- ✓ moles of H_2O_2 in the sample
- ✓ mass of H_2O_2 in the sample
- ✓ percent by mass of H_2O_2

1. Use your data and the concentration provided for the permanganate solution to calculate the mass of solution, and the number of moles of $KMnO_4$ used in each of your titrations. Show your work for the first titration. Enter values for all three titrations in the Summary Table.

2. Use your results for **1** and the balanced equation given for the reaction between potassium permanganate and hydrogen peroxide to determine the mass and number of moles of hydrogen peroxide that were present in each sample. Show your calculations for the first sample, and enter the values for all three in the Summary Table.

3. Use the mass of H_2O_2 (from **2**) and the mass of the H_2O_2 solution to determine the mass percent of H_2O_2 in each sample. Show your work for the first trial and enter all three trials in the Summary Table.

4. Calculate the average deviation for your three trials. Report the average mass percent of H_2O_2 with average deviation.

5. Discuss the sources of experimental error for this experiment. Remember that errors in calculation and incorrect procedure are not experimental errors.

6. Consider the original question: "Does a commercial antiseptic have the full potency claimed on the label?" You've completed your analysis: does it have the strength it is supposed to have? (**Hint:** Notice the number of significant figures claimed by the manufacturer.)

Something Extra

1. Find a bottle of hydrogen peroxide for which the expiration date has passed. With your teacher's permission, devise and carry out experiments to determine how much potency it has lost.

2. Exposure to sunlight and heat both tend to hasten the decomposition of hydrogen peroxide. With your teacher's permission design and run experiments to measure the effects of these variables.

Experiment 82

Galvanic Cells

Problem

How can we measure the electrical energy that is produced in oxidation-reduction reactions?

Introduction

Recall that an oxidation-reduction reaction involves a transfer of electrons. In this experiment we will consider a device for transferring electrons from a reducing agent to an oxidizing agent through a wire, rather than by direct contact in the solution. For example, consider the situation shown in **Figure 1a** where zinc metal is in contact with Zn^{2+} ions in one beaker and copper metal is in contact with Cu^{2+} ions in another beaker. Both solutions also contain anions which are only spectator ions (do not participate in the oxidation-reduction reaction). Because zinc gives up electrons more easily than copper, electrons would tend to flow from the left beaker to the right beaker. However, so far there is no connection between the beakers which will allow electron flow.

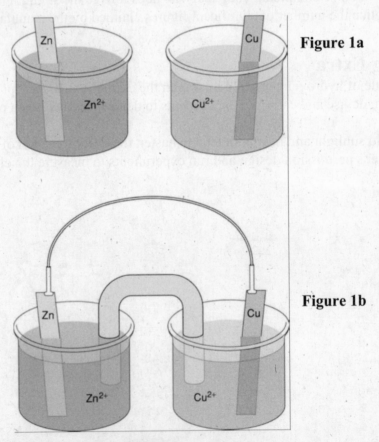

Figure 1a

Figure 1b

Figure 1

First the zinc metal must be connected to the copper metal by a wire as shown in **Figure 1b.**
However, if electrons flow from zinc metal to copper metal this would leave the left beaker with a
positive charge and would produce a negative charge in the right beaker. Therefore, to prevent this
charge buildup we also need a way for ions to flow. This is achieved by using a **salt bridge**. Your
text explains that a salt bridge contains a solution of a strong electrolyte, usually in a gel so that the
solution can't run out. As electrons are transferred from zinc to copper , Zn^{2+} ions form in the left
beaker and copper ions accept electrons in the right beaker and become copper atoms. Thus,
because the number of cations increases in the left-hand chamber, anions must leave the salt bridge
to balance the charge. Similarly, as copper(II) cations are removed from the right-hand chamber,
they are replaced by cations from the salt bridge.

The two chambers, joined by the salt bridge and the wire, constitute an electrochemical cell,
commonly known as a **galvanic cell**, in honor of Luigi Galvani, an Italian scientist who is generally
credited with discovering electricity.

Note that oxidation is occurring in the left-hand chamber (referred to as the **anode**), and copper(II)
ions are being reduced in right-hand chamber (the **cathode**). That is:

Oxidation occurs at the anode.

Reduction occurs at the cathode.

Experimental Design

Read this section carefully, and be sure you understand it, before you
begin the actual experiment.

In your experiment, you will use a simple setup to construct galvanic cells. The beakers are replaced
by the wells of a 24-well test plate, and a piece of soft cotton string, soaked in a sodium nitrate
solution, serves as the salt bridge. The electrodes are small strips of the various metals or pieces of
metal wire. The metal electrodes are in contact with solutions of their most stable cation. **Figure 2**
is a diagram of the setup.

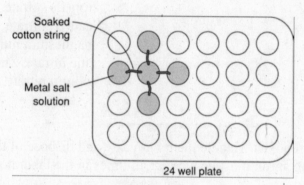

Figure 2
Set-up for measuring cell potentials

Using this arrangement, you can run two sets of experiments at once or repeat the experiment if you
wish to confirm your results. With copper at the center of the five-well set in the upper left, its
reduction potential can be compared with those of four other metals. Each well contains a 1.0 M
solution of one of the metal cations, with a small piece of the metal sticking out of it. The short,

dark lines in the figure represent the string salt bridges. The five metals you will use are copper, magnesium, iron, lead, and zinc. The copper(II) ions will take electrons from each of the other metals, so it will be the reference point. The CBL voltage probe consists of two leads with small clips that can be extended to hook on to the small metal electrodes.

Note 1: In order to obtain valid readings, it is essential that the voltage probes be in contact with the metal pieces only. If the probes dip into the solutions you will get inaccurate results.

Note 2: The two leads of the voltage probe should always be connected so that the reading on the CBL is positive.

Prelaboratory Assignment

✓ Read the **Introduction** and **Procedure** before you begin.
✓ Answer the Prelaboratory Questions.

1. In part 1 of the procedure, is the Cu^{2+}/Cu well the cathode or the anode in each case? Explain.

2. Given the nature of the reactions that take place in part 1, what might you expect to find if you were to weigh the copper electrode before the experiment and again after completion? Why?

Materials

Apparatus
Graphing calculator (TI 83+, or other)
CBL, CBL 2, or LabPro
DataMate or ChemBio program
CBL voltage probe
24-well test plate
Soft cotton string or filter paper strips
Sandpaper or steel wool
Forceps
Safety goggles
Lab apron

Reagents
Small strips or short wire lengths of:
copper, Cu
iron, Fe
lead, Pb
magnesium, Mg
zinc, Zn
1.0 M solutions of the cations:
copper(II) sulfate, $CuSO_4$
iron(II) sulfate, $FeSO_4$
lead(II) nitrate, $Pb(NO_3)_2$
magnesium nitrate, $Mg(NO_3)_2$
zinc nitrate, $Zn(NO_3)_2$
sodium nitrate, $NaNO_3$

Safety

1. Some of the reagents are toxic. Handle them with care and dispose of them properly.
2. Safety goggles and a lab apron must be worn at all times in the laboratory.

Procedure

If you are not using a Report Sheet, prepare a Data Table for Part 1 with the following headings:

Cell Combination	Observed Potential	Rank vs. Copper

1. Use your connecting cable to attach your graphing calculator to the CBL, CBL2, or LabPro unit. Attach the voltage probe, using Channel 1 of the CBL unit. If you are using the original CBL and a TI 83-plus graphing calculator, open ChemBio from APPS. With CBL2 or LabPro, open DataMate. If the calculator displays Voltage (V) in CH 1, all is ready. If not, set up the unit for one voltage probe in channel 1.

Part 1: Relative potentials *vs.* copper

2. Using a clean 24-well test plate, set up the five wells as shown in **Figure 2**. Place enough copper sulfate solution in the center well to make it about 1/3 full. Place similar amounts of the other four cation solutions in the perimeter wells. Be sure to record which solution is in which well.

3. Soak four pieces of soft cotton string in a small amount of sodium nitrate solution for a moment, then place the pieces so that each connects the center well to one of the perimeter wells.

4. Place a piece of copper in the center well and pieces of each of the other metals in the well that contains the cation of that metal. **Note:** The pieces of metal should not be in direct contact with the string salt bridge.

5. Connect one probe to the piece of copper and the other to one of the other metals. Be sure that the probes are only in contact with the metals, not the solutions. The voltage reading should be positive; if it is not, reverse the probes. Note and record the voltage reading. **Note:** The accepted method for identifying cells is to put the anode first, then the cathode. Since copper metal is the cathode in each case in Part 1, it always comes second, e.g., Mg/Cu.

6. Leaving the copper piece connected, move the other lead successively to each of the other metals. Note and record the voltage for each case on the Data Table.

7. Now answer questions 1 and 2 of **Analysis and Conclusions**. Your answers here form a sort of hypothesis for Part 2 of the experiment.

Part 2: Relative potentials of the perimeter metals.

There are six possible combinations that you can make from the four perimeter metals. As you carry out Part 2 of the procedure, enter the observed values in the Data Table for Part 2.

1. Remove the salt bridges from Part 1. Discard them as directed by your teacher.

2. For this part of the experiment, you will compare magnesium, zinc, iron and lead to each other. While there are a number of ways to accomplish this goal, you might consider the one illustrated in **Figure 3**.

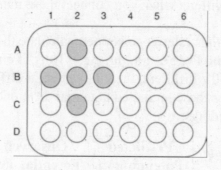

Figure 3
Example of a well plate for determining relative reduction potentials

The rows and columns are numbered as they are on the 24-well plate and the same wells are shaded as in **Figure 2**. Use fresh segments of string soaked in sodium nitrate solution to connect the pairs you want to test. Be sure that each well being used for a particular test has only one piece of string in it--the one connecting the two metals being tested.

In well C3, make a half-cell identical to the one in B1. You can now connect B3 to A2, C2, and C3. You can also connect C3 and C2. The only combination you don't have at this point is between B1 and B3. You can solve this by making a half-cell in D3 that is identical to A2. This will give you the complete set of six pairings called for in Part 2.

Cleaning Up

1. Your teacher will instruct you as to how the CBL units and voltage probes are to be put away. Be sure you clean the probe clips with distilled water.
2. The bits of metal are not soluble in water, so they cannot go down the drain. Because some of the metal solutions are toxic, they must not get into the water supply. Shake the contents of your well plate into the large tray lined with paper towels. Once the liquid has dried, the towels and residue can be safely incinerated.
3. Rinse the well plate, shake it dry, and put it away.
4. Dispose of the salt bridge strings as directed by your teacher. Excess sodium nitrate solution may be rinsed down the drain with large amounts of water.
5. Wash your hands thoroughly with soap and water before leaving the laboratory.

Analysis and Conclusions

Complete the **Analysis and Conclusions** section for this experiment either on your Report Sheet or in your lab report as directed by your teacher.

1. On the Data Table for Part 1 rank the four perimeter metals in terms of their relative difference to copper, with 1 assigned the metal with the largest potential difference.

2. In Part 1 of the experiment, you determined the reduction potentials for four metals, all relative to copper. Now use these potentials and the ranks that you assigned the four metals to predict the potentials you will get when you connect these metals(in pairs) directly to each other.
   a. By comparing their potentials relative to copper, calculate the potential difference for each of the six pairs (Fe and Pb, Fe and Mg, etc.) Thus, if one metal is 0.75V different than copper, and another is 0.50V different, then the potential difference between the two must be 0.75 - 0.50 = 0.25V. Enter these predicted potential differences in a Data Table for Part 2 that has the following headings:

Metal Pair	Anode/Cathode	Predicted Potential (V)	Observed Potential (V)	Deviation (V)	Percent Deviation

   b. Cations of all of the four metals are more easily reduced than Cu^{2+}. That is, in combination with each of the four, copper is always the cathode. By comparing their relative positions versus copper, decide which member of each pair should be the anode (oxidized) and which should be the cathode (reduced). Enter your answers in the Data Table for Part 2

3. Based on your results, which lead of the voltage probe (red or black) should be connected to the anode and which to the cathode?

4. Determine the difference between each of your predictions and the values you measured in Part 2. Then divide each deviation by the measured value and convert the resulting fraction to a percent. Show one sample calculation and enter that and all other results in the Data Table for Part 2.

5. Determine your experimental values for the standard reduction potentials of each of the other four metals by adding -0.34V to the observed potentials relative to copper.

6. Consult an advanced text or a standard reference, such as the CRC *Handbook of Chemistry and Physics* to find the accepted values for iron, lead, magnesium and zinc. Calculate your percent error for each of the values and enter it in the table above. In each case, the percent error is found by dividing the difference between your value and the accepted value, by the accepted value and then convert to percent.

Experiment 83

Corrosion of Iron

Problem
What are some of the factors that cause corrosion of metals? What can be done to minimize this phenomenon?

Introduction
Corrosion is a general term applied to the process in which metals are converted to oxides or to other compounds. Iron (III) oxide is commonly called rust. Silver sulfide is another common product of corrosion; it is the compound that appears as tarnish. The process of corrosion gradually deteriorates the metals involved. Although the detailed chemistry of the corrosion is often not completely understood, it clearly involves oxidation and reduction. Most metals are easily oxidized; that is, they lose electrons relatively easily. Conversely, the oxygen gas in the atmosphere is a good oxidizing agent because atoms of oxygen easily gain electrons (are reduced).

During this investigation you will study some of the factors involved in the corrosion of iron. You will observe the behavior of iron in the presence of a number of solutions as well as in the presence of other metals. From your experimental results, you will try to see patterns or make generalizations about the process of corrosion.

Prelaboratory Assignment
✓ Read the **Introduction** and **Procedure** before you begin.
✓ Answer the Prelaboratory Questions.

1. Determine the oxidation state of the iron atom in each of the following compounds:
   a. FeO   **b.** Fe_2O_3

2. Why must oxidation and reduction always occur simultaneously? In other words, why will there never be an oxidation reaction without an accompanying reduction?

3. What environmental conditions might accelerate the rate at which a metal corrodes? Why?

4. Metals are known to have relatively low ionization energies. How does this fact relate to the ease with which metals corrode?

Materials

Apparatus
9 small iron finishing nails
6 small test tubes (13x100 mm)
150-mL beaker
test tube rack
pH or litmus paper test strips
Petri dish

Reagents
Set of reagent solutions
Phenolphthalein
Steel wool
Copper wire
Zinc strip or mossy zinc
Prepared agar solution

Pliers
Stirring rod
Disposable pipets
Safety goggles
Lab apron

0.1 M potassium hexacyanoferrate solution, $K_3Fe(CN)_6$
0.2 M iron(II) sulfate solution, $FeSO_4$

Safety

1. Wear safety goggles and a lab apron at all times in the laboratory.
2. No food or drink is allowed in the laboratory at any time.

Procedure

If you are not using a Report Sheet for this experiment design a Data Table for Part 1 with the following headings:

Test Tube	Reagent	Acid/Base/Neutral	Observations of Reaction

Part 1   Reactions of Iron with Aqueous Reagent Solutions

1. Place a clean, bright nail into each of five test tubes. Use steel wool if necessary to clean the surface of the nails. **Note:** Slide each nail carefully down the side of the test tube to avoid breaking the bottom of the test tube.

2. Obtain a set of reagent solutions from your teacher (other teams will do the two other groups). Partially fill each of the test tubes with one of the reagents in the solution set so that the nail is just covered.

Group A	Group B	Group C
NaOH	KOH	Na_3PO_4
$Na_2Cr_2O_7$	Na_2CO_3	$Na_2C_2O_4$
NaCl	KNO_3	KSCN
HCl	HNO_3	H_2SO_4
distilled water	distilled water	distilled water

3. Dip pieces of pH or litmus paper into each solution to determine whether the solution is acidic, basic or neutral.

4. Allow the nails to stand overnight in the solutions and proceed to Part 2 of the procedure.

Part 2   Reactions of Iron with Metals

1. Obtain about 100 mL of prepared agar solution in a small beaker.

2. Add about 10 drops of 0.1 M $K_3Fe(CN)_6$ and 10 drops of 1% phenolphthalein indicator solution to the agar mixture. Stir thoroughly.

3. Place a clean, bright nail in a Petri dish. Bend another clean nail sharply with a pair of pliers and place it in the same dish. **Note:** Be sure that the nails do not touch each other.

4. Twist a clean piece of bare copper wire around a third nail. Temporarily remove the nail from the coil and tighten the wire coil so that when the nail is forced back through, it makes tight contact with the wire coil. Place this nail in the lid of the Petri dish. See **Figure 1**

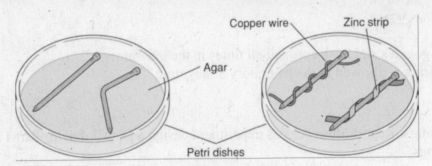

Figure 1
Placement of nails in Petri dishes containing agar

5. Twist a clean piece of zinc around a fourth nail. As in the previous step, be sure that the zinc makes tight contact with the nail. If a strip of zinc is not available, a piece of mossy zinc may be attached by forcing the nail through the zinc piece in at least two places. Place this nail also in the lid of the Petri dish. **Note:** Be sure that the nails do not touch each other. See **Figure 1**.

6. Carefully pour the agar solution into the Petri dishes until the nails and attached metals are covered to a depth of about 0.5 cm.

7. Make observations during the time remaining in the class period. Place the dishes in a safe place as directed by your teacher and allow them to sit overnight.

Part 3   Final Observations of the Reactions

1. At the start of the next class period, make observations of the five test tubes containing the nails and reagent solutions. Record on any changes that have take place.

2. To each of the five solutions add 1-2 drops of $0.1 M\ K_3Fe(CN)_6$ (which contains K^+ and $Fe(CN)_6^-$). Observe any changes which occur and record your results.

3. In a separate test tube, add 1 drop of $0.1 M\ K_3Fe(CN)_6$ to about 1 mL of $FeSO_4$ solution. Compare this result to that obtained when the $K_3Fe(CN)_6$ was added to the test tubes in Step 2.

4. Now make observations of the reactions in the Petri dishes. Sketch a picture of what you observe. Be sure to indicate colors that you notice.

Cleaning Up

1. Dispose of the nails and agar mixture in the solid waste container.
2. Clean out the test tubes as directed by your teacher. Wash all glassware and return it to its proper location.
3. Wash your hands before leaving the laboratory.

Analysis and Conclusions

Complete the **Analysis and Conclusions** section for this experiment either on your Report Sheet or in your lab report as directed by your teacher.

1. List the reagents in Part 1 for which there was no reaction. In addition to your own data, report the findings of at least one other lab group which used the same set of reagents you did.

2. List the reagents in Part 1 for which there was indication of corrosion. In addition to your own data, report the findings of at least one other lab group which used the same set of reagents you did.

3. Are there any regularities among the reagents which caused corrosion? What evidence did you observe?

4. In Parts 2 and 3, what reactions did you observe at the head of the nail, at the pointed end, and at the sharp bend of the nail? Were these reactions different from those for the remainder of the nail? How do you explain this?

5. Iron(II) ions are known to react with the hexacyanoferrate ion of potassium hexacyanoferrate to form a colored precipitate. Write an equation for this reaction.

6. In Part 2, what colors appeared in the agar in the Petri dishes? What does each color indicate?

7. What color is indicates oxidation? What color is indicates reduction?

8. How does a coating of zinc on iron protect iron from corrosion?

9. Why can a nail be stored on the shelf and not rust, but when it is placed in water, it rusts quickly?

Something Extra

1. Consult a reduction potential chart and predict another metal that is more readily oxidized than iron.

2. Using chemical resources investigate the process known as galvanization. How does it relate to your experiment?

Experiment 84

Investigating Radioactivity

Problem
How is the intensity of radiation affected by distance from the source?

Introduction
Radioactive decay results from instability in the nucleus of the atoms involved. The nucleus contains protons and neutrons. Recall that the nucleus occupies only a tiny fraction (about one one-quadrillionth (10^{-15}) of the volume of the atom. Given its small size it would seem that the repulsion of the positive charges of the protons should cause the nucleus to fly apart. It appears that one function of the neutrons is to help somehow hold the protons together, acting as a sort of "nuclear glue." When the ratio of neutrons to protons in the nucleus does not provide the necessary stability, its nucleus spontaneously decomposes, releasing energy in the form of photons and particles. Chapter 19 explains that there are a number of types of naturally-occurring radiation. The three best known are alpha, beta, and gamma, but there is also positron emission. *Alpha particles* are by far the largest of the radiation particles, and do not travel well through the air. As a result, the procedure for studying alpha radiation will be somewhat different than it is for beta and gamma radiation.

Beta particles are high-speed electrons. The net effect of beta emission is to convert one of the neutrons in the atom's nucleus into a proton and an electron. Positron radiation is similar, but in this case a proton appears to be converted to a neutron and the nucleus emits a particle that is identical in mass to an electron, but it has a positive charge.

Gamma radiation consists of a stream of photons -- bundles of energy. Gamma radiation generally accompanies one of the particle forms of radiation. Gamma rays have no effect on the atomic number or mass number of the isotope undergoing decay.

In this experiment, you will work with one or more radioactive sources, alpha, beta, or gamma. These sources involve extremely small amounts of radiation, so they present no health hazard. Your main task will be to investigate how distance from the source affects the intensity of radiation. By combining your results with those of your classmates, you will learn whether the effect of distance is the same for all three types of radiation. The Vernier Radiation Monitor uses a Geiger-Mueller tube, of the type described in Chapter 19. It has an audio counter and a flashing count light; you will use the silent counter here, since the CBL is doing the counting for you. Your first step will be to measure the background radiation of the laboratory for later comparison with your experimental results.

Prelaboratory Assignment

✓ Read the **Introduction** and **Procedure** before you begin.
✓ Answer the Prelaboratory Questions.

1. How would the neutron:proton ratio be affected by:
   a. beta radiation? Defend your answer.
   b. positron production? Defend your answer.

2. A proton has a mass of 1.67262×10^{-27} kg. The mass of an electron is 9.10939×10^{-31} kg.
   a. What is the combined mass of a proton and an electron? Watch your precision!
   b. If a neutron has a mass of 1.67493×10^{-27} kg, how much mass is "lost" during beta decay?
   c. If that mass is lost as a photon of electromagnetic radiation, use the equation, $E = mc^2$, to calculate the amount of energy produced in the process.

Materials

Apparatus
TI 83+, or similar graphing calculator
CBL, CBL-2, or Lab Pro
Vernier Radiation Monitor with cable
Link cable
CBL-P adapter
Meter stick
Safety goggles
Lab apron

Reagents
Radioactive source: alpha (α), beta (β), or gamma (γ)

Safety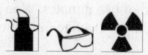

1. The radiation sources present no health hazard, but common-sense laboratory practice should always be observed.
2. Safety goggles and a lab apron must be worn at all times in the laboratory.

Procedure

In order to use the radiation monitor, you will need to install the program, RADIATIN, in your calculator. This program may be downloaded from the Vernier Scientific website, at www.vernier.com. Use Graph Link cable to load the program from your computer to your calculator. Once the program has been installed on one calculator, it can be quickly transferred to others. Perhaps your teacher has already downloaded and installed the program on her or his calculator. Unlike most of the programs used with the TI83+, RADIATIN is not part of the ChemBio or Physics applications found in APPS; once it has been installed, you will find it in PRGM (programs).

Setting Up the System

Connect the calculator to the CBL using the link cable, then plug the CBL-P adapter into Channel 1 on the top of the CBL unit. The large plug on the connector that comes with the Radiation Monitor is plugged into the CBL-P adapter; the small, 1/8 inch end goes to the Radiation Monitor. See **Figure 1**.

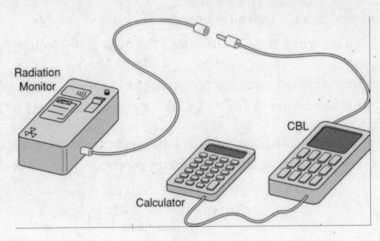

Figure 1
CBL assembly for measuring radiation

Part 1   Background Radiation

1. Be sure that all radioactive sources have been removed by at least 4 meters from where you are working.

2. Turn on the calculator, the CBL and the Radiation Monitor. Start the RADIATIN program and proceed to the MAIN MENU. Select COLLECT DATA and BACK-GROUND. The calculator and CBL will then measure the background radiation for a period of two minutes. You will be given the option to subtract this amount from all subsequent readings. Decline the option, but record the value on the Report Sheet.

Part 2   Effect of Distance on Radiation Intensity

Note: The distances given below (steps 2-4) are for beta and gamma sources, but will not work for alpha radiation, due to its restricted ability to move through the air. If you are using an alpha source, replace the 5-, 10-, 15-, 20-, and 25-**cm** distances with 0-, 2-, 4-, 6-, 8-, and 10-**mm** distances. The distance of 0.0 mm requires the source to be placed flush with the face of the monitor, and the monitor is placed against the end of the meter stick.

1. Return to the MAIN MENU of the RADIATIN program, again select COLLECT DATA. Select TRIGGER/PROMPT. **See Figure 2.**

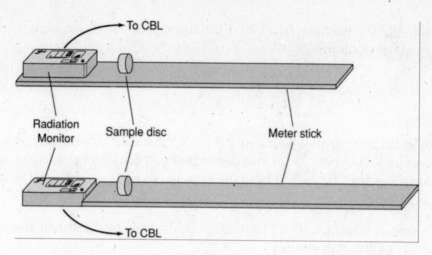

Figure 2
Placement of radiation sources and monitor
Top: Placement for alpha source
Bottom: Placement for beta or gamma source

2. Place the monitor in the appropriate position for the type of radiation being used, as shown in **Figure 2**. Place the radioactive source on the meter stick at a distance of 5.0 cm from the monitor. If one side of the sample disk has a metal foil label, place the source so that the foil label faces **away from** the radiation monitor.

3. Begin the collection by pushing ENTER on your calculator. When the Monitor has completed its count, the screen will show PROMPT?; enter 5. You will then have the option to either collect more data or stop and view what you have. Accept the first option (COLLECT MORE DATA?).

4. Reposition the radioactive source at 10.0 cm from the Monitor and repeat the data collection. Continue in this fashion, increasing the distance to 15.0-, 20.0-, and finally 25.0-cm. Once the 25.0 cm reading has been completed, accept the option to stop and view the graph. The calculator screen will tell you where the data are stored (usuallyL1 and L2). Press ENTER to view a plot of your data. You can get values for the individual points either by using TRACE or by going to the lists themselves (STAT, EDIT).

Cleaning Up

1. Return all apparatus and radioactive sources to their proper places.
2. Wash your hands thoroughly before leaving the laboratory.

Analysis and Conclusions

Complete the **Analysis and Conclusions** section for this experiment either on your Report Sheet or in your lab report as directed by your teacher.

1. Using either the TRACE function, or STAT/EDIT from your calculator, complete the table below (cpm = counts per minute).

Distance	5.0 cm	10.0 cm	15.0 cm	20.0 cm	25.0 cm
Intensity (cpm)	_____	_____	_____	_____	_____

2. Use a piece of graph paper to plot your data of cpm *vs.* distance. If available, use a French curve to make the best smooth-curve. Otherwise do the best you can, freehand, using the screen of your calculator as a guide. Be sure to label the axes and to use scales that fill as much of the graph paper as possible.

3. Describe the shape of your plot. Does increasing the distance by consistent amounts cause the radiation intensity to diminish steadily?

4. Use your graph to estimate the distance at which the radiation from the source is not distinguishable from the background radiation that you measured in **Part 1**.

5. **a.** Calculate the ratio of the intensities for the following pairs of distances.

    10 cm/5 cm   15 cm/10 cm   20 cm/15 cm

   b. What can you deduce from the way in which the ratios change?

6. Compare your graph with those produced by teams that used other types of radiation. Are the shapes the same? If so, is the rate of decrease comparable for all three forms of radiation?

7. Radiation, like light or sound, spreads out as it leaves its source; think about what you see when you drop a pebble in a still pool of water. The sketch below illustrates this spreading effect.

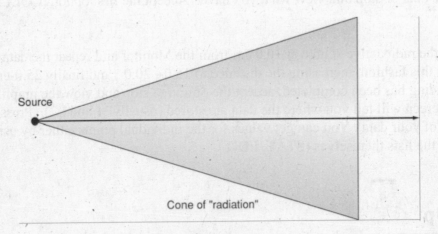

Source

Cone of "radiation"

Figure 1

Mark the center line off into segments, such that 2 cm on the line represents a 5-cm distance from the source. For this purpose, we will assume that the sensor on the radiation monitor can be represented by an object 2 mm across. Place a 2-mm hash-mark at each 2-cm interval, and calculate the fraction of the cone's width that 2-mm represents at that distance. Describe what you find. How does the fraction of the radiation that actually hits the target change as distance increases? This change in intensity is called *attenuation*.

8. The simulation in Question 7 uses a flat cone to represent the manner in which radiation spreads out. In fact, it spreads out in a full 360° fashion, unless restricted by shielding. How would this change the rate of attenuation?

Something Extra

1. If your radioactive source has a reasonably short half-life (a few minutes, up to about an hour), use the HALF-LIFE option of the Monitor and compare your result with the accepted value for the half-life of the isotope you're testing.

2. With your teacher's approval, design and carry out experiments to investigate various types of shielding, first with paper, then with aluminum foil and lead foil. **[Note: lead metal is toxic; use gloves when handling the lead foil. Wash your hands thoroughly after the experiment.]** Consider the following questions:
   - How does the intensity of radiation vary with the type of shielding used?
   - Which type of shielding is the most effective on all three types of radiation?
   - Does shielding affect all types of radiation in the same way, or is one type of radiation easier to stop than others?

Experiment 85

The Half-Life of Pennies

Problem
What does flipping pennies have to do with the concept of half-life?

Introduction
The half-life of a radioactive sample is the time required for half of the original sample of nuclei to decay. Knowing the half-life of carbon-14, for example, enables us to determine the age of wooden artifacts.

Prelaboratory Assignment
✓ Read the entire experiment before you begin.
✓ Answer the Prelaboratory Questions.

1. In this experiment, what do the pennies that land "heads" represent?
2. In this experiment, what do the pennies that land "tails" represent?
3. In this experiment, what does each flip represent?

Materials
100 pennies
Graph paper

Procedure
1. Flip 100 pennies and separate them according to which landed heads and which landed tails. Record the number of heads.

2. Flip only the pennies that landed heads, and then separate the pennies according to which landed heads and which landed tails. Record the number of heads. Repeat this until you are out of pennies. Record the number of times until you are out of pennies.

Cleaning Up
1. Leave the pennies on the lab bench.
2. Wash your hands thoroughly before leaving the laboratory.

Analysis and Conclusions

Complete the **Analysis and Conclusions** section for this experiment either on your Report Sheet or in your lab report as directed by your teacher.

1. Make a graph of number of pennies flipped vs. trial number from your data.

2. Gather together all of the class data and make a second graph of the total number of pennies flipped vs. trial number.

3. Why is there a difference between the graph of your data and graph of the class data?

4. Draw a graph that shows the decay of a 100.0-g sample of a radioactive nuclide with a half-life of 10 years. This should be a graph of mass versus time for the first four half-lives.

5. Compare the two graphs using your data and the class data to the graph of the 100.0 g sample. Does your graph or the graph of the class data look more like the graph of the 100.0-g sample? Why?

6. Approximately how many half-lives would it take for one mole of a radioactive nuclide to completely disappear?

Something Extra

Would the shape of the graph change if you used a different number of pennies? Try this activity again with a different number of pennies and comment on the results. Use a wide range (from 10 pennies to a few hundred pennies).

Experiment 86

Synthesis of Esters

Problem
What simple organic compounds can be synthesized in the laboratory?

Introduction
One of the most important commercial (and natural) types of organic syntheses involves the reaction of an organic acid, called a *carboxylic acid*, with an alcohol to produce an ester. A simple example, shown below, is the reaction between methyl alcohol, CH_3OH, and acetic acid, CH_3COOH. The reaction is presented with structural formulas to better illustrate the way the reaction takes place, although the actual mechanism is of no interest to us here.

$$CH_3-C\overset{O}{\underset{OH}{\big\langle}} \quad + \quad HO-CH_3 \quad \overset{H^+}{\longrightarrow} \quad CH_3-\overset{O}{\overset{\|}{C}}-O-CH_3 \quad + \quad H_2O$$

The H^+ above the reaction arrow indicates that the reaction requires an acid catalyst. Esters are generally characterized by sweet, often pleasant odors, including many of the familiar fruit aromas. Banana oil, for example, is the result of ester formation, produced by the reaction between acetic acid (shown in the reaction above) and isoamyl alcohol, $CH_3CH_2CH(OH)CH_2CH_3$.

In this experiment, you will make one or more esters, carrying out the reaction on a small scale. Since you are only interested in the nature of the product, and have no concern for the stoichiometry or percentage yield, precise measurements of reactant quantities is not necessary. All that you need to do is to combine the reactants, then allow them to boil gently together for a short time. This process, in which a reaction mixture is maintained at a slow boil, without a loss of material, is known as *refluxing*. A *condenser* attached to the top of the reaction vessel allows any evaporating liquid to cool, recondense, and return to the liquid in the tube.

While the goal of the experiment is to make the specific ester methyl salicylate, your teacher may provide the reactants for you to make one or more esters, since the process is quick and simple to carry out. These may be assigned as part of the activity, or they may be offered as extra credit.

Prelaboratory Assignment
✓ Read the **Introduction** and **Procedure** before you begin.
✓ Answer the Prelaboratory Questions.
1. Explain how the reflux condenser helps to keep the liquid reagents from boiling away during the esterification reaction.

2. Using the equation for the reaction between methanol and acetic acid as a model, write the equation for the reaction between methanol and salicylic acid. Consult Chapter 20 of your text for the structure of salicylic acid.

3. Suggest an explanation for the fact that esterification reactions are often referred to as *condensation reactions*.

Materials

Apparatus
Test tube, 13 x 100 mm
Pasteur pipet
Hot plate with heat transfer block
Parafilm™ or other laboratory film
    strips, about 1 cm x 5 cm
Boiling chips
Wire test tube clamp
Safety goggles
Lab apron
Gloves (optional)

Reagents
Salicylic acid, crystals
Methanol (methyl alcohol)
Concentrated (18M) sulfuric acid, H_2SO_4

Safety

1. Concentrated sulfuric acid is extremely corrosive to skin and clothing. Handle it with great respect, use it only in the fume hood, and use only the dropper that comes with the dispensing bottle. Gloves may be worn to protect your hands when pouring the concentrated acid.
2. The reaction mixtures contain sulfuric acid. Neutralize any spills and clean them up with large amounts of water.
3. Methanol is highly flammable, as are all of the alcohols in the Optional Reagents list. Be sure there are no open flames in the laboratory.
4. Follow safe lab technique when testing the aroma of your product. The odors of some of the esters, although familiar and pleasant, are quite penetrating and harsh in large doses.
5. Safety goggles and a lab apron must be worn at all times in the laboratory.
6. Remember that hot plates and hot glassware look the same as cool objects. Take care to avoid burns.

Procedure

If you are not using a Report Sheet for this experiment be sure to include the following in your Data/Observations:
    Initial Appearance and odor of:
      ✓ salicylic acid
      ✓ methanol

Describe what you saw during the refluxing. Could you tell that liquid was being recondensed? If so, what did you see that indicated this was happening?
    Appearance and odor of:
      ✓ methyl salicylate
      ✓ water

1. Make a reflux condenser by wrapping a small piece of Parafilm™ or other laboratory film around the wide end of a Pasteur pipet. Use enough to allow the pipet to fit snugly into the top of a standard 100-mm test tube as shown in **Figure 1**. Wrapping two or three layers around the pipet usually is enough. Set the condenser aside for a moment. Turn the hotplate on, using the low-power setting.

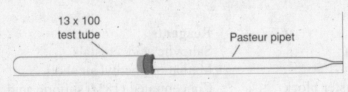

13 x 100 test tube   Pasteur pipet

Figure 1
Reflux assembly

2. Place a small amount of salicylic acid (2-hydroxybenzoic acid, $C_6H_4(OH)COOH$) in the tube; use enough to give a depth of about 0.5 cm of the solid acid.

3. Add an approximately equal volume of methyl alcohol; the acid will dissolve in the alcohol, although perhaps not completely at first.

4. Go to the fume hood where you will find a small dropper bottle of concentrated sulfuric acid; this is the catalyst. **DANGER: sulfuric acid is very harmful.** Using the dropper provided, add 2 or 3 drops of the acid to your reaction mixture, then return the dropper to the bottle. Finally add a boiling chip to the tube, taking care to avoid splashing.

5. Insert the Pasteur pipet/reflux condenser into the top of the reaction tube. Using a wire test tube holder, allow the tube to rest lightly on the surface of the hot plate, in the small hole in the heat transfer block. The mixture should begin to reflux within a minute or two.

   You can control the rate at which the refluxing takes place by raising the tube off the surface of the hot plate if boiling becomes too rapid, or by lowering it if the reaction slows below the desired rate.

   Maintain the reflux for two or three minutes, then remove the tube to a beaker or other heat-proof holder and allow it to cool for a few minutes.

6. Carefully remove the condenser, then hold the tube about 6 inches away from your face, and wave your hand across the top of the tube to direct the odor of your product toward your nose. The chemical name of the ester you produced is *methyl salicylate*; you may recognize it as oil of wintergreen.
   Note: If you are going to carry out one or more additional esterification reactions, clean up the first one as described in the following section, *except* do not discard the reflux condenser; it can be used again, so long as the Parafilm™ does not deteriorate (If it does, replace the film).

Cleaning Up

1. Methyl salicylate is a naturally-occurring substance, as are the other esters listed under **Something Extra**. They can be rinsed down the drain, using a large amount of water to dilute the residual sulfuric acid. (Remember that the sulfuric acid acts a catalyst, so it was not consumed during the reaction. Therefore, whatever you put in the tube is still there.)
2. The boiling chip must not be allowed to enter the drain; it should be recovered, rinsed, and placed in the container for non-hazardous solid waste.
3. Once you are through with all of the esterifications that you plan to do, place the Pasteur pipet in the container labeled BROKEN GLASS ONLY.
4. Wash your hands before leaving the laboratory.

Analysis and Conclusions

Complete the **Analysis and Conclusions** section for this experiment either on your Report Sheet or in your lab report as directed by your teacher.

1. In this experiment you were interested only in the nature of the products. Explain how you would revise the experiment if the goal was to determine the percent yield. Explain completely what you would do and what measurements would be needed.

Something Extra

Any number of familiar fragrances can be made from simple, common alcohols and acids, but many of the more familiar ones involve acids or alcohols that have offensive or even toxic properties. Notable are those involving butyric acid, an extremely unpleasant smelling carboxylic acid, resulting in fragrances such as apple and pineapple.

With your teacher's permission try some of the combinations of carboxylic acids and alcohols listed below. Do not create your own combinations without consulting your teacher. The fragrances of some common esters are listed. However, even those for which no fragrance is given will be recognizable.

The procedure for these combinations is the same as before: about 0.5 cm depth of acid, an equal volume of alcohol, and 2-3 drops of sulfuric acid catalyst are refluxed gently for several minutes. Disposal is to be carried out as described under **Cleaning Up**.

Carboxylic Acid	Alcohol	Fragrance
acetic acid	ethyl alcohol (ethanol)	----------
acetic acid	amyl alcohol (pentyl alcohol)	banana
acetic acid	octyl alcohol	orange
acetic acid	n-propanol	pear
benzoic acid	ethyl alcohol	----------
formic acid	isobutanol	raspberry
propionic acid	isobutanol	rum

Experiment 87

Saponification

Problem

How is soap made? How does soap clean soiled articles? How does soap made in the laboratory compare to commercially-prepared soap?

Introduction

Soap is something that we regularly buy in the grocery store and use daily. This was not always the case. Queen Isabella of Spain (1451 - 1504) claimed that she had bathed twice in her entire life, once when she was born and again on her wedding day. Queen Elizabeth I of England (1558 - 1603) took a bath every three months whether she "needeth it or no". Cleopatra, the beautiful queen of Egypt bathed in fragrant oils. The oils softened her skin, and the perfumes were needed to camouflage the odors produced by the bacteria on her skin. Soap was known in ancient Rome – a soap factory was revealed among the ruins of ancient Pompeii – but this soap was too harsh to use on skin.

The use of soap and water as we know it began in London when people realized that poor personal hygiene was part of the cause of cholera and typhoid epidemics. In 1846, the British government passed a Public Baths and Wash Act. It provided public baths and laundries for the working class of London. The idea rapidly spread throughout Europe and the United States.

Soap is produced when a fat (or oil) is mixed and heated with sodium hydroxide, commonly known as lye, in a chemical process called saponification. Solid soap is the sodium salt of a fatty acid. Soft soap is a mixture of soap and glycerol. Liquid soap is the potassium salt of a fatty acid.

In this experiment you will synthesize soap and then analyze its properties through several different chemical tests.

Prelaboratory Assignment

✓ Read the **Introduction** and **Procedure** before you begin.
✓ Answer the Prelaboratory Questions.

1. You will prepare a sodium hydroxide solution in this experiment. If 8.0 grams of NaOH are dissolved in enough water to make 25.0 mL of solution, what is the molarity of that solution?

2. Why are many organic molecules nonpolar and, therefore, not water soluble? What property of a soap molecule gives it an affinity for water?

Materials

Apparatus
Spatula or spoon
250-mL beaker
10-mL graduated cylinder
Heating unit
Stirring rod
3 small test tubes (13 x 100mm) and stoppers
Test tube rack
Safety goggles
Lab apron
Centigram or milligram balance
Disposable pipets

Reagents
Oil or fat (choose one)
   Cottonseed oil
   Coconut oil
   Tallow
   Lard
   Vegetable shortening
Ethyl Alcohol
Solid NaOH pellets
Phenolphthalein
Commercial soap
Commercial detergent
0.5 M CaCl$_2$ solution

Safety

1. Wear safety goggles and a lab apron at all times in the laboratory.
2. No food or drink is allowed in the laboratory at any time.
3. Be very careful when heating the soap mixture and especially when adding the ethyl alcohol. Ethyl alcohol is flammable.
4. Sodium hydroxide is caustic. Be careful not to get it on your skin or clothing.

Procedure
Part 1   Making the Soap
1. Weigh out 15-20 g of your selected oil or fat. Transfer it to a clean 250 mL beaker.

2. To the beaker add 7-8 grams of solid sodium hydroxide pellets and 25 mL of tap water.

3. Measure 10 mL of ethyl alcohol using a graduated cylinder. Add this to the 250 mL beaker containing the fat and NaOH solution.

4. Heat the mixture carefully on the heating unit, stirring it **constantly** for at least 20 minutes. The solution should be hot but not boiling. As you heat carefully drip in 10 mL of additional ethyl alcohol. **Note:** If you are using a flame for heating be careful to keep the alcohol away from the flame. As the mixture is heated, it will bubble and foam. Be careful to regulate the burner by moving it as necessary to prevent the mixture from overflowing.

5. At the end of 20 minutes, add 25 mL of water and continue heating and stirring for an additional 10 minutes. Allow the mixture to cool.

6. Remove the coagulated soap mass with a spoon or spatula. Place it on several layers of paper towel and pat it dry. Using the towel, shape it into a bar or a ball. Observe and record its appearance and smell.

7. Allow the soap to air dry until the next laboratory session.

Part 2   Testing the Soap

1. Place 5 mL of water in each of three test tubes. Add a small portion of your soap to the first test tube. Stopper the tube and shake vigorously to test the foaming action of your soap. Repeat this process using commercial soap in the second test tube and commercial detergent in the third test tube. Record your observations. Clean out the test tubes.

2. Place 5 mL of water in each of three test tubes. Add 8-10 drops of vegetable oil to each tube. Note that the oil forms a separate layer on top of the water. Stopper and shake the first tube. Allow the tube to stand for a few minutes and observe again. Add a small portion of your soap to the first test tube. Stopper the tube and shake. Allow the tube to stand a few minutes. Record your observations. Repeat this process using commercial soap in the second test tube and commercial detergent in the third test tube. Record your observations. Clean out the test tubes.

3. Place 5 mL of water in each of three test tubes. Add a small portion of your soap to the first test tube. Stopper the tube and shake to dissolve the soap. Add 10-15 drops of $CaCl_2$ solution to the soap solution. Shake and record the results. Repeat this process using commercial soap in the second test tube and commercial detergent in the third test tube. Record your observations. Clean out the test tubes.

4. Place 5 mL of water in each of three test tubes. Add a small portion of your soap to the first test tube. Stopper the tube and shake to dissolve the soap. Add 2-3 drops of phenolphthalein indicator solution. Shake and record your observations. Repeat this process using commercial soap in the second test tube and commercial detergent in the third test tube. Record your observations. Clean out the test tubes.

Cleaning Up

1. Dispose of the soap in the solid waste container.
2. Clean out the test tubes as directed by your teacher. Wash all glassware and return it to its proper location.
3. Wash your hands thoroughly before leaving the laboratory.

Analysis and Conclusions

Complete the **Analysis and Conclusions** section for this experiment either on your Report Sheet or in your lab report as directed by your teacher.

1. How do the sudsing actions of the soaps and detergent compare?

2. How do the oil breakup capabilities of the soaps and detergent compare?

3. Calcium chloride releases calcium ions into the water. This makes the water "hard". How do the results of the soaps and detergent compare in hard water?

4. How do the pH values of the commercial soap and detergent compare with your soap?

Something Extra

1. Check the labels of commercial soaps and detergents. Determine how they are different. What makes a detergent a better choice for use in hard water?

2. Making homemade soap can also be done using lard and ashes from a fire. The ashes of burned wood contain sodium and potassium oxides. What happens when these oxides react with water?

Experiment 88

Synthesis of Slime

Problem
How is slime made and what are some of its properties?

Introduction
Undoubtedly you have encountered "slime," perhaps as a toy or in a grade school science demonstration. It is fun to play with, but it also has much to teach us about charge distributions, the nature of intermolecular forces, polymers, and the cross-linking of polymer chains.

The commercial product known as Slime is produced by the Mattel Toy Corporation and is slightly different from the material we will synthesize. Our slime is produced by mixing a 4% solution of polyvinyl alcohol with a mixture of boric acid, a very weak acid, and sodium hydroxide, a strong base. They react to form sodium tetraborate, ordinary borax. Borax acts as a cross-linker, holding the long, stringy polyvinyl alcohol molecules together making a slimy semisolid gel-like substance that can be picked up, stretched, broken, molded, and bounced, even though it is 96% water. Cross-linking gives the network stability, but when subjected to a shearing stress it can be broken down. This polymer is quite soluble in a large excess of water and it tends to turn fluid under stress.

Polyvinyl alcohol contains a long chain of carbon atoms with hydroxyl groups on every other carbon atom. It is a linear polymer and can have a molar mass of up to 150,000 g/mol with almost 7000 carbon atoms in a row. Structures of vinyl alcohol and polyvinyl alcohol are shown below. Polyvinyl alcohol is made by connecting hundreds of polyvinyl alcohol molecules. The double line connecting the carbon atoms of vinyl alcohol represents two pairs of electrons being shared between the same two atoms. As you will learn, such bonds are common in organic molecules and they tend to be more reactive than single bonds. Notice that the polymer no longer has the double bonds; they break during polymerization, becoming single bonds.

$$H-\overset{\displaystyle H}{\underset{|}{C}}=\overset{\displaystyle H}{\underset{|}{C}}-OH$$

Vinyl Alcohol

Polyvinyl alcohol

Ordinarily a substance with such a large molar mass would not be soluble in water, but the large number of hydroxyl groups that can form *hydrogen bonds* with water contribute strongly to the remarkable solubility of polyvinyl alcohol. Hydrogen bonds are not true chemical bonds. Rather, they are one of the forces of attraction that hold one molecule close to another. These are called intermolecular forces (*inter* means between). In fact, hydrogen bonds are the strongest type of these intermolecular forces. They result when the hydrogen atom on one molecule carries a partial positive charge and the oxygen atom in the other molecule carries a partial negative charge and a pair of unshared electrons. You first encountered these interactions in connection with the strong attractions that water molecules have for each other.

Hydrogen bonds are broken as the temperature of the system rises, causing molecular motions to increase. What would be the expected effect of heating slime?

The slime will be prepared by simply stirring a solution of sodium borate into a solution of polyvinyl alcohol. In a minute or two, the mixture becomes quite thick; and in about 10 minutes it sets to a firm gel, if the conditions are proper. In your experiment, the sodium borate will be prepared by mixing boric acid with sodium hydroxide; this is an acid-base neutralization reaction, and the salt produced is sodium borate. What must be the other product?

Prelaboratory Assignment

✓ Read the **Introduction** and **Procedure** before you begin.
✓ Answer the Prelaboratory Questions.

1. Hydrogen bonds are broken as the temperature of the system rises. What would be the expected effect of heating slime?

2. What makes a large molecule such as polyvinyl alcohol soluble in water?

3. Write a balanced molecular equation for the formation of sodium borate and water from boric acid and sodium hydroxide. Formulas are given in Materials.

Materials

Apparatus
Paper cups, small
10-mL graduated cylinder
pH paper or Universal Indicator
Wood applicator sticks
1-mL graduated pipet
Safety goggles
Lab apron

Reagents
4% Mass/volume aqueous polyvinyl alcohol solution
0.2 M Aqueous boric acid solution, H_3BO_3
0.2 M Aqueous sodium hydroxide solution, NaOH
Sodium silicate solution, saturated (water glass)

Safety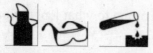

1. Take extra precautions handling sodium hydroxide and sodium silicate solutions. They are both extremely caustic.
2. While slime prepared from sodium tetraborate (borax) is not toxic or corrosive, it is very difficult to remove from certain fabrics, including rugs and carpets.
3. Safety goggles and a lab apron must be worn at all times in the laboratory.

Procedure

Part 1   Making slime

For each step of the procedure, briefly describe **what you did and what you saw**.

1. Place 2 mL of 0.2 M boric acid solution in a paper cup. Then measure the pH of the solution by using a wood stick to touch a drop of it to a piece of pH indicator paper. Add 1 mL of 0.2 M

NaOH and test the pH again using a different spot on the piece of pH indicator paper. Add another 1 mL of NaOH and test again. Finally, add a third 1-mL portion of NaOH and test the pH a final time. At this point, you should have a mixture of 2 mL of boric acid and 3 mL of NaOH. Set the mixture aside for use in step 6.

2. In a wax-paper cup place 10 mL of 4% polyvinyl alcohol solution and add 2 mL of 0.2 M boric acid solution. Stir the mixture thoroughly. A wood stick works very well for this. Note the appearance and viscosity of the mixture.

3. Add 1 mL of 0.2 M sodium hydroxide solution to the polyvinyl alcohol-boric acid mixture. Stir well and observe the viscosity of the mixture.

4. Add a second 1-mL portion of 0.2 M sodium hydroxide solution. Stir the mixture well and again note the nature of the mixture. Has the viscosity changed? How?

5. Add a third 1-mL portion of NaOH to the mixture in the cup, stir again, and note any changes in the appearance and viscosity of the mixture.

6. Pour 10 mL of 4% polyvinyl alcohol solution into a fresh cup and add the mixture of 2 mL of 0.2 M boric acid and 3 mL of 0.2 M sodium hydroxide solution that you made in step 1. Stir this combination thoroughly and note the nature of the mixture. Does it seem different in appearance, texture or viscosity from the material you obtained at the end of step 5? Can you tell whether premixing of the boric acid and sodium hydroxide makes any difference in the properties of the final product?

7. Remove the material from the first cup, place it in a small beaker and heat it on a steam bath or hot plate for 2 to 3 minutes, with stirring. Record any changes in the physical properties of the hot mixture?

Part 2   Properties of Slime

Carry out the following tests of your product. Record your observations on your Report Sheet or in your notebook.

1. After cooling to room temperature, remove the mixture from the beaker. Roll the slime into a ball and place on a flat surface. What happens to it over a period of time?

2. Hit the relaxed ball of slime hard with your hand. What happens to it?

3. Pick it up and stretch it slowly. Note what happens. Try to determine to what length you can stretch it without having it break.

4. Roll it into a cylinder and pull on it rapidly. Note what happens.

5. Can you bounce a ball of slime? What happens when you try it? How hard can you drop it?

6. Try dissolving a small piece of slime in about 25 mL of water. Stir the mixture thoroughly.

7. In a third cup place 2 mL of 0.2 M boric acid solution and 3 mL of 0.2 M sodium hydroxide solution. To this mixture add about a gram of sodium chloride and stir the mixture thoroughly to dissolve as much of the salt as possible. Add 10 mL of 4% polyvinyl alcohol solution and stir the mixture. Compare the results with the material from Part 1 steps 5 and 6.

Cleaning Up

1. The rigid and semi-rigid polymers can be disposed of in the non-hazardous waste container (waste basket).
2. Small quantities of unused liquid solutions can be washed down the drain with lots of water.
3. Wash your hands thoroughly before leaving the laboratory.

Analysis and Conclusions

Complete the **Analysis and Conclusions** section for this experiment either on your Report Sheet or in your lab report as directed by your teacher.

1. Summarize your observations. Use complete sentences with correct spelling and grammar.

2. Describe the effect of sodium borate on the polyvinyl alcohol solution and summarize the properties you discovered for slime.

3. What difference(s) in the properties of slime did you notice between the product of step 5 and those of the product from step 6?

Something Extra

The polymers you have investigated used sodium borate to cross-link the long polyvinyl alcohol chains together. But borate ion is not the only substance that can do this cross-linking. Another polymer can be made using sodium silicate in place of sodium borate. Sodium silicate has the empirical formula, Na_2SiO_3, but it is, in fact a polymeric chain:

Repeating SiO_3^{2-} Unit

The Silicate Structure

Note the similarities between this structural formula and the one for polyvinyl alcohol shown in the Introduction. Notice that there is a 1:3 ratio between silicon and oxygen.

How might the properties of this polymer differ from those you have observed so far? With your teacher's permission, carry out the following procedure to find out.

1. In a paper cup, mix 2 mL of sodium silicate solution and 10 mL of 4% polyvinyl alcohol solution. Stir the mixture for about a minute and collect the product on the end of the wood stick. Rinse the product briefly with water in a beaker, then squeeze out excess water between paper towels. Compare the properties of this material with the slime you made in the experiment.

2. Try mixing equal volumes of sodium silicate solution with the polyvinyl alcohol solution. Label your cups so that you can note any change in the properties of this material a day or so after it has been prepared.

Experiment 89

Gluep

Problem
Can you make a cross-linked polymer from white glue?

Introduction
Polymers are large chain-like molecules that are built from small molecules called monomers. You have used polymers, such as Teflon and nylon, in everyday life. Polymers are also used in plastic bottles, carpets, clothing, and synthetic rubber.

Cross-linked polymers consist of polymer chains connected by chemical bonds. Substances such as Slime or Gak are cross-linked polymers. Cross-linking polyvinyl acetate (present in most white glues) with laundry borax can make a substance called gluep.

In this lab activity you will make gluep from white glue and laundry borax. You will also test its properties.

Prelaboratory Assignment
✓ Read the entire experiment before you begin.

Materials

Apparatus	Reagents
Safety goggles	Water
Lab apron	Borax solution
Plastic cups	White glue
Wax paper	Baking Soda
Plastic spoons	Vinegar
Zip-lock bag	Food coloring

Safety

1. Safety goggles and a lab apron must be worn at all times in the laboratory.
2. If you come in contact with any solution, wash the contacted area thoroughly.

Procedure
Part 1   Making Gluep
1. Pour about 15 mL (one tablespoon) of white glue into a plastic cup.

2. Add about 15 mL of water to the white glue and stir. Add a drop or two of food coloring. Make and record careful observations.

3. Add about 10 mL (two teaspoons) of borax solution (provided by your teacher) to the glue and water mixture and stir. Make and record careful observations.

4. Use the plastic spoon to transfer the gluep onto wax paper. Use a paper towel to get rid of excess liquid.

5. Squeeze out the excess liquid and roll the gluep into a ball.

6. Write a paragraph discussing your observations when making gluep.

Part 2   Testing Gluep's Properties

1. Does a ball made from gluep bounce? Drop your ball from a height of 2 feet onto a non-carpeted floor. Measure and record how high the ball bounces.

2. Some substances you can buy at a toy store can lift an image from the newspaper. Can gluep do this?

Part 3   Reacting Gluep with Vinegar and Baking Soda

1. Place a small ball (1 cm diameter) of gluep in a cup. Add about 20 drops of vinegar and stir. Record your observations.

2. Add about half a teaspoon of baking soda to this mixture and stir. Record your observations.

Cleaning Up

1. Clean up all materials and return them to their proper locations.
2. Dispose of all extra chemicals as instructed by your teacher.
3. You may keep your gluep if you like. Be sure to put it in a zip-lock bag.
4. Wash your hands thoroughly before leaving the laboratory.

Analysis and Conclusions

Complete the **Analysis and Conclusions** section for this experiment either on your Report Sheet or in your lab report as directed by your teacher.

1. Write a paragraph summarizing the properties of gluep.

2. Explain your observations when gluep reacted with vinegar and baking soda.

Something Extra

Vary the proportions of water and borax solution in making gluep and retest its properties.

Experiment 90

Enzymes in Food

Problem
What is the effect of a type of enzyme called a protease on gelatin? How can we affect the functioning of proteases?

Introduction
Enzymes are proteins that act as catalysts for specific biological reactions. Without enzymes life would be impossible because important reactions in our bodies would occur far too slowly.

Enzymes called proteases break down certain proteins. Since some fruits contain proteases and gelatin is a network of protein molecules, we can use gelatin to experiment on the functioning of proteases.

Prelaboratory Assignment
✓ Read the entire experiment before you begin.
✓ Answer the Prelaboratory Question.
   1. Why is there one sample of gelatin to which nothing was added?

Materials

Apparatus
Safety goggles
Lab apron
Plastic cups
Toothpicks
Microwave oven

Reagents
Frozen pineapple
Fresh pineapple
Canned pineapple
Meat tenderizer
Soft contact lens cleaner
Dry gelatin

Safety
1. Safety goggles and a lab apron must be worn at all times in the laboratory.

Procedure
Part 1   Preparing the gelatin
1. Prepare the gelatin according to the directions on the package.

2. Divide the gelatin equally into seven plastic cups. Label the cups and place them in the refrigerator.

3. Allow the gelatin to set.

Part 2   How do proteases affect gelatin?

1. Place a piece of fresh pineapple (about a 1-inch cube) in a microwave oven and cook on the highest setting for one minute. Allow the pineapple to return to room temperature.

2. Test the effect of proteases on gelatin by placing the following on the surface of the gelatin in each cup:

    Cup 1:   nothing
    Cup 2:   a piece (1-inch cube) of fresh pineapple
    Cup 3:   a piece (1-inch cube) of canned pineapple
    Cup 4:   a piece (1-inch cube) of frozen pineapple
    Cup 5:   the piece of pineapple that was microwaved
    Cup 6:   a spoonful of meat tenderizer
    Cup 7:   a teaspoon of contact lens cleaner

3. Examine the cups every 5 minutes for 20 minutes. Be careful to return the pieces of pineapple to their original positions after looking under each piece. Record all observations.

4. Return the cups to the refrigerator.

5. Observe the cups the next day and record your observations.

Cleaning Up

1. Dispose of your cups in a plastic lined trashcan.
2. Return all materials to their proper locations.
3. Wash your hands thoroughly before leaving the laboratory.

Analysis and Conclusions

Complete the **Analysis and Conclusions** section for this experiment either on your Report Sheet or in your lab report as directed by your teacher.

1. Why do many packages of gelatins state "do not use fresh or frozen pineapple"? Why do the packages not include the statement "do not used canned pineapple?"

2. Is canned pineapple more like fresh, frozen, or microwaved pineapple? What does this tell you about the canning process? Explain.

3. From your results, would you say meat tenderizers contain proteases?

4. From your results, would you say contact lens cleaners contain proteases?

5. Propose a theory of meat tenderizes and contact lens cleaners work.

Something Extra

How do proteases affect the setting of gelatin? Carry out an experiment to answer this question.

Experiment 91

Vitamin C in Juices

Problem
How can we determine which juices provide the full MRDA (minimum required daily allowance) of vitamin C?

Introduction
Fruits and juices are a primary source of vitamin C in the diet. But are all fruits equal? Your goal in this investigation is to answer that question for yourself. To do this, you and your lab partner will test a number of juices, fruit drinks, and other beverages with the intent of determining their value as a source of vitamin C. Recognize that you are taking a very narrow measure of the juices, and that a low vitamin C content does not mean that the juice has no other food value, nor does a high vitamin C content necessarily make one juice overall superior to the others. Which would you be more inclined to favor for your own diet: an unpleasant tasting liquid that was high in vitamin C, or a pleasant tasting one that has half the vitamin C content?

As you may remember from biology or another science class, iodine is used to test for the presence of starch. When a deep red iodine solution is added to a material that contains starch, the red color is changed to a dark blue, sometimes almost black shade. Vitamin C (ascorbic acid) will also react with iodine solution, but it does so in such a way that the red color of the iodine is lost. If an iodine solution is added to a solution containing both vitamin C and starch, the iodine will react with the vitamin C first, and the iodine color will disappear; then, when the vitamin C has all been consumed, any additional iodine that is added will react with the starch to produce the dark blue color.

This will be the basis for your analysis. You will place a little starch in each of several beverages, then you will add iodine solution, one drop at a time, until the solution turns blue, indicating that all of the vitamin C has been consumed. If you always use the same amount of beverage, you can compare the vitamin C contents of the different products. And, if you carry out the same procedure on a solution whose vitamin C content is accurately known, you can actually determine the amount of vitamin C in a serving of that beverage.

This is a cooperative venture between you and your lab partner. This experiment will be more interesting if you can solve a real-world problem. Perhaps you could test the effect of leaving a beverage container open for long periods of time; or you might look for a connection between the percentage of real fruit juice and vitamin C content. Your teacher will help you set a goal.

A total of ten juice drinks is to be analyzed and compared with a Standard Vitamin C solution. Each of you is to test the Standard, then analyze five of the juices, so you will do a total of six analyses each.

Prelaboratory Assignment

✓ Read the **Introduction** and **Procedure** before you begin.
✓ Answer the Prelaboratory Questions.

1. What is the purpose of the standard vitamin C solution?

2. The vitamin C standard must be freshly prepared and must be kept chilled in order to maintain the proper concentration of vitamin C.

   a. What does this tell you about the effect on vitamin C content of leaving beverage containers open for long periods of time?

   b. How could you experimentally test your hypothesis?

3. In two separate lists, identify the juices that you will test and those that your partner will test.

Materials

Apparatus
Small beakers, flasks or other titration vessels (3 or more)
Safety goggles
Lab apron

Reagents
Standard vitamin C solution in thin-stem pipet, 1 mg/mL
Iodine solution, in microtip pipet
Starch indicator solution, in thin-stem pipet
Beverages to be tested
Sodium thiosulfate solution for disposing of excess iodine solution

Safety

1. Wear laboratory safety goggles at all times in the laboratory.
2. Some individuals are sensitive to iodine, and it can stain skin and clothing. Handle it with care.
3. Even though this experiment involves pleasant tasting food products, you must never eat or drink anything in the laboratory. **Nor should you take any of the items for later consumption.**

Procedure

Part 1   Standardization of the Iodine Solution.

1. Place exactly 25 drops of Vitamin C Standard solution in a small test tube, beaker, or other vessel and add 2 drops of starch solution. Record the concentration of your standard in your observations.

2. Add Iodine Test solution, one drop at a time, agitating the vessel gently after each drop to help with mixing. Continue until the mixture turns blue, indicating that all of the vitamin C has been consumed by iodine. (The blue color may be somewhat faint.) Record the number of drops of iodine needed.

Part 2   Analysis of beverages

1. Carry out the same procedure as in Part 1, but use 25 drops of your five juices in place of the 25 drops of Vitamin C Standard. As before, add two drops of starch, then add the iodine solution dropwise. In each case, record the number of drops of iodine needed to achieve a blue color.

   Note: Some of the beverages have colors of their own, so you will need to use some judgement in deciding when you have reached the endpoints of the analyses.

Cleaning Up

1. The small amount of iodine remaining in the test tubes is not significant, and the juice drinks are not hazardous, so all test tubes may be cleaned in the sink. Similarly, remaining Vitamin C Standard may be rinsed down the drain.

2. Iodine Test solution should not be put down the drain. Empty your iodine pipet into a small container, then add sodium thiosulfate solution, one drop at a time, until the color of the iodine disappears. The solution is now safe to flush down the drain with water.

3. Wash your hands thoroughly with soap and water before leaving the laboratory.

Analysis and Conclusions

Complete the **Analysis and Conclusions** section for this experiment either on your Report Sheet or in your lab report as directed by your teacher. If you are not using a Report Sheet prepare a Summary Table with the following headings:

(a) Juice	(b) Drops of Iodine	(c) Conc. of Ratio to Standard	(d) Conc. of Standard (mg/mL)	(e) mg Vitamin C in Juice (mg/mL)	mg Vitamin C in 180 mL of juice

In general, fruit juices and fruit drinks are less concentrated, so will require fewer drops of iodine than the standard did, although that is not always the case. By comparing the number of drops of iodine needed for a given beverage with the number needed for the same amount (25 drops) of the Standard, you can get a relative strength comparison. If you then use the actual concentration of the Vitamin C Test solution, you can determine the actual vitamin C concentration in each juice in units of milligrams of Vitamin C per milliliter (mg/mL). Finally, by multiplying this concentration by the number of milliliters in a typical serving, and then comparing your answer with the MRDA for vitamin C, you can decide how much of that particular drink you would need to satisfy your body's need for Vitamin C.

You will do all the calculations for the standard and the five juices that you tested. Your partner will do the same. Later, you will combine your results with those of your partner to see what conclusions you can draw.

1. Calculate the relative concentration of vitamin C in each juice: divide the number of drops of iodine needed for 25 drops of juice by the number of drops needed for 25 drops of Standard. Express the ratio as a decimal with two significant figures. Show your calculations and enter your results in column (b) of the Summary Table.

2. Enter the concentration of the vitamin C standard in column (c) of the Summary Table. This value is the same for all trials. The concentration is found on the stock bottle of vitamin C standard.

3. Find the concentration of vitamin C in each juice by multiplying the ratios you calculated in **1**, by the concentration of vitamin C in the standard. Enter the results in column (d) of the table.

4. We will assume that a standard serving of beverage is six fluid ounces (6 fl. oz.), or about 180 mL. Multiply the concentration of each beverage, in mg/mL, by this volume to determine the number of milligrams of vitamin C in one serving. Show your calculations and enter the results in column (e) of your table.

5. The average high school student, male or female, needs about 60 mg of vitamin C per day (this is called the **minimum recommended daily allowance** (MRDA). Consulting with your partner, list the ten juices and drinks in descending order of vitamin C content. In case of ties, use alphabetical order. Make your list in two vertical columns, with the first column containing those that deliver at least the MDRA of vitamin C in one 180-mL serving, and with those that do not in the second column.

6. For each of the juices in your second column (the ones that do not deliver adequate vitamin C in one serving), calculate the volume in milliliters that you would have to drink in order to achieve the 60 mg MDRA.

7. Do any of the juices surprise you, either by how well or how poorly they tested? In particular, were there any that you thought would perform better than they did?

8. Based on the results of this experiment, and on your own personal preferences, what juice would be the best one to get your 60 mg of vitamin C?

9. While some of the beverages had color, we did not test cola drinks, grape juice or grape drink, or prune juice. Suggest a reason why these would not be suitable for this experiment.

Something Extra

1. Consult the library or some other reference source to find some sources of vitamin C other than fruits and juices. List two or three of the best ones. How could you test the vitamin C content of a fresh fruit or vegetable that was not in the form of a juice?

Appendix A

Vapor Pressure of Water at Various Temperatures

Temperature (°C)	Pressure (mm Hg)	Temperature (°C)	Pressure (mm Hg)
0	4.580	31	33.70
5	6.543	32	35.66
10	9.209	34	39.90
15	12.79	36	44.56
16	13.63	38	46.49
17	14.53	40	55.32
18	15.48	45	71.88
19	16.48	50	92.51
20	17.54	55	118.0
21	18.65	60	149.4
22	19.83	65	187.6
23	21.07	70	233.7
24	22.38	75	289.1
25	23.76	80	355.1
26	25.21	85	433.6
27	26.74	90	525.8
28	28.35	95	633.9
29	30.04	100	760.0
30	31.82		

1A

1
H
Hydrogen
1.008

2A

3	4
Li	**Be**
Lithium	Beryllium
6.941	9.012

11	12
Na	**Mg**
Sodium	Magnesium
22.99	24.31

Key

6	— Atomic number
C	— Element symbol
Carbon	— Element name
12.01	— Atomic mass

Period 1: H

Period 2: Li, Be

Period 3: Na, Mg

	19	20	21	22	23	24	25	26	27
4	**K**	**Ca**	**Sc**	**Ti**	**V**	**Cr**	**Mn**	**Fe**	**Co**
	Potassium	Calcium	Scandium	Titanium	Vanadium	Chromium	Manganese	Iron	Cobalt
	39.10	40.08	44.96	47.88	50.94	52.00	54.94	55.85	58.93
	37	38	39	40	41	42	43	44	45
5	**Rb**	**Sr**	**Y**	**Zr**	**Nb**	**Mo**	**Tc**	**Ru**	**Rh**
	Rubidium	Strontium	Yttrium	Zirconium	Niobium	Molybdenum	Technetium	Ruthenium	Rhodium
	85.47	87.62	88.91	91.22	92.91	95.94	(98)	101.1	102.9
	55	56	57	72	73	74	75	76	77
6	**Cs**	**Ba**	**La**	**Hf**	**Ta**	**W**	**Re**	**Os**	**Ir**
	Cesium	Barium	Lanthanum	Hafnium	Tantulum	Tungsten	Rhenium	Osmium	Iridium
	132.9	137.3	138.9	178.5	180.9	183.9	186.2	190.2	192.2
	87	88	89	104	105	106	107	108	109
7	**Fr**	**Ra**	**Ac**	**Rf**	**Db**	**Sg**	**Bh**	**Hs**	**Mt**
	Francium	Radium	Actinium	Rutherfordium	Dubnium	Seaborgium	Bohrium	Hassium	Meitnerium
	(223)	226.0	(227)	(261)	(262)	(263)	(264)	(265)	(268)

58	59	60	61	62
Ce	**Pr**	**Nd**	**Pm**	**Sm**
Cerium	Praseodymium	Neodymium	Promethium	Samarium
140.1	140.9	144.2	(145)	150.4

90	91	92	93	94
Th	**Pa**	**U**	**Np**	**Pu**
Thorium	Protactinium	Uranium	Neptunium	Plutonium
232.0	(231)	238.0	(237)	(244)

8A

3A	4A	5A	6A	7A	8A
					2 He Helium 4.003
5 B Boron 10.81	6 C Carbon 12.01	7 N Nitrogen 14.01	8 O Oxygen 16.00	9 F Fluorine 19.00	10 Ne Neon 20.18
13 Al Aluminum 26.98	14 Si Silicon 28.09	15 P Phosphorus 30.97	16 S Sulfur 32.07	17 Cl Chlorine 35.45	18 Ar Argon 39.95

28 Ni Nickel 58.69	29 Cu Copper 63.55	30 Zn Zinc 65.38	31 Ga Gallium 69.72	32 Ge Germanium 72.59	33 As Arsenic 74.92	34 Se Selenium 78.96	35 Br Bromine 79.90	36 Kr Krypton 83.80
46 Pd Palladium 106.4	47 Ag Silver 107.9	48 Cd Cadmium 112.4	49 In Indium 114.8	50 Sn Tin 118.7	51 Sb Antimony 121.8	52 Te Tellurium 127.6	53 I Iodine 126.9	54 Xe Xenon 131.3
78 Pt Platinum 195.1	79 Au Gold 197.0	80 Hg Mercury 200.6	81 Tl Thallium 204.4	82 Pb Lead 207.2	83 Bi Bismuth 209.0	84 Po Polonium (209)	85 At Astatine (210)	86 Rn Radon (222)
110 Ds Darmstadtium (271)	111 Rg Roentgenium (272)	112 Uub Ununbium (277)	113 Uut Ununtrium (284)	114 Uuq Ununquadium (289)	115 Uup Ununpentium (288)			

63 Eu Europium 152.0	64 Gd Gadolinium 157.3	65 Tb Terbium 158.9	66 Dy Dysprosium 162.5	67 Ho Holmium 164.9	68 Er Erbium 167.3	69 Tm Thulium 168.9	70 Yb Ytterbium 173.0	71 Lu Lutetium 175.0
95 Am Americium (243)	96 Cm Curium (247)	97 Bk Berkelium (247)	98 Cf Californium (251)	99 Es Einsteinium (252)	100 Fm Fermium (257)	101 Md Mendelevium (258)	102 No Nobelium (259)	103 Lr Lawrencium (260)